**Concepts for
Operations Management**

Concepts for Operations Management

Ray Wild

Administrative Staff College, Henley and
Brunel University, UK

A Wiley–Interscience Publication

JOHN WILEY & SONS

Chichester · New York · Brisbane · Toronto

Library of Congress Cataloging in Publication Data:
Wild, Ray.
Concepts for operations management.

'A Wiley–Interscience publication.'
Includes index.
1. Industrial management. 2. System analysis.
I. Title. II. Title: Operations management.

HD31.W5157   658.4   77–7232

ISBN 0 471 99539 8 (cloth)
ISBN 0 471 99543 6 (paper)

Printed by William Clowes and Sons, Limited
London, Beccles and Colchester.   78-6381

Contents

CHAPTER 1

Introduction

The following chapters explore some aspects of operations management. They seek to set out some simple basic concepts for operations management, and in so doing they aim to help the reader understand the *nature* of operating systems and operations management. They do not aim to tell the reader how to manage operating systems. This is not a handbook. It has little to do with problem solving and is not intended to replace other such operations management books. It provides an introduction to the subject, a foundation on which to build. It will provide insufficient information for the intending practitioner, but hopefully will help him organize his thoughts and use the other books and information which are available to him.

The approach adopted here is different but not new. The ideas contained in the following chapters owe something to previous work in other areas, but are applied here for the first time in this context. They are simple, perhaps over simple, and there is a limit to the extent to which they can usefully be applied. Like any simple concepts or models, any attempt to apply them beyond a certain level may lead to confusion rather than clarification. The approach taken is largely descriptive, but only simple descriptions are attempted, for the principal intention is to provide a basis for understanding rather than tools and techniques for practice. We shall refer to techniques and procedures—not in detail, but simply to provide a bridge or link between our approach and the treatments of the subject offered elsewhere.

These chapters aim to help fill a gap. They relate to the second stage of three stages in the development of operations managers, i.e.

(1) Exposure.
(2) Concepts.
(3) Action.

The purpose of these three stages is to provide some first hand familiarity with operating systems and the activities of operations management, to provide an understanding of the nature, significance and problems of operating systems and management, and to develop an individual's ability to manage operations.

Whilst study or work at one stage may facilitate or reinforce learning at another, these stages should be tackled substantially in this order, and in particular in operations management an understanding of concepts is a prerequisite

1

to action. Some may not need to accomplish all stages. Those destined for jobs in operations management will need to satisfy all three whilst those intended for other functions will require stage 2 plus perhaps some appreciation of stage 3.

Stage 2 must deal with theory whilst stage 3 is concerned with decision making, problem solving, techniques, and practice. It seems that education for operations management has gone wrong, or at least become confused in the treatment of these two stages. Whilst in other fields higher education emphasizes concepts or theory, the teaching of operations management seems primarily to be concerned with techniques and practice. The major part of most courses and texts is oriented to action, and the prerequisite of concepts or theory is largely neglected. In many cases therefore it seems that those individuals destined for operations management are required to concentrate upon action or practice without having achieved the necessary prior understanding of concepts and theory, and that those seeking only a familiarity with operations management are exposed, inappropriately, to action considerations and thus fail to achieve an adequate understanding of the nature of operating systems and operations management. For these reasons some teaching of operations management may have been ineffective. Students may leave management courses ill equipped to deal with the management of operations. They might also consider their courses to be inadequately based upon an intellectual foundation, insufficiently supported by a unified or coherent body of knowledge and therefore rather uninspiring, unattractive and dull.

Whilst most existing operations management books deal explicitly with stage 3, there have been attempts to tackle the subject from a concepts and theory viewpoint. Case study work provides one means to approach stage 2, although in many cases there is a prerequirement to study problems, decisions, etc (i.e. stage 3). Other approaches to the examination of concepts and theory in this area have not as yet met with much success. The 'techniques' approach does not seek to tackle stage 2, but deals explicitly with stage 3. The 'life-cycle' approach (which emphasizes the chronological order in which operations management problems occur) provides an appropriate framework for the examination of operations management problems but in itself sheds little direct light on fundamental concepts. The 'systems' view promised much but seems to have provided little.

Apart from their failure adequately to tackle concepts, some previous treatment of operations management seems to have 'scope deficiencies'. Much of what is written on operations management seems to be largely about manufacturing management. We have tried to avoid this pitfall in our rather simple approach to the subject. A definition of operations management is developed and much of the book is devoted to the clarification of the scope and nature of operating systems.

The approach adopted is designed to shed some light on the nature of operating systems, and the nature, responsibilities and role of the operations manager. Naturally the operations manager is seen as a key central figure in management. We nevertheless accept that in too many cases he, and his function, is inadequately represented at a policy level in the organization. There is little that a book can do to change the status and rewards of operations management, but attempt to deal with the subject so that its role vis-à-vis other functions, and its importance is highlighted.

To recapitulate this is an attempt to provide a NEW, SIMPLE, DESCRIPTIVE treatment of OPERATING SYSTEMS and OPERATIONS MANAGEMENT—an attempt to identify basic concepts for operations management.

USING THIS BOOK FOR TEACHING OPERATIONS MANAGEMENT

This book is intended to help those who wish to think about the nature of operating systems and operations management. It attempts to *develop* ideas and concepts and therefore it may not be appropriate for readers simply to 'dip into it'. It probably has to be taken as a whole. The ideas presented are simple but novel, and for this reason they will not meet with unanimous approval. They will hopefully provoke thought, and certainly some thought will be necessary in reading the book.

It might be appropriate for students to consider the issues raised in this book before examining the *practice* of operations management. Some or all of the contents may therefore provide an adequate introduction to operations management. It may be beneficial also to use some other supplementary material in adopting this approach to teaching operations management. Comments and advice on the use of this book in teaching, further teaching material, suggested assignments, readings, exam questions, etc, can be obtained from the author.

ACKNOWLEDGEMENTS

Numerous colleagues were kind enough to comment on earlier drafts of this material (including Dr N D Slack, Dr C C Gallagher, Professor T A J Nicholson, Dr D W Birchall and Dr C A Carnall). Many useful suggestions were offered. In particular I am indebted to Dr G M Buxey for several valuable ideas and comments. Mr A V King and Mr M Jennings suggested the mnemonic introduced in Chapter 4, and the 'building development case' material presented in Chapter 14 is based on material developed by Mr M Jennings. I am indebted to Mrs F Stubbs for repeatedly typing this manuscript.

CHAPTER 2

The Nature and Relevance of Operations Management

Perhaps the toughest problem in attempting to deal with the subject of operations management is knowing how best to start. To some extent we must overcome a semantic problem. We must agree, or agree to disagree on definitions. More importantly we must agree on the scope of our subject. We can tackle both problems simultaneously, and perhaps the best starting point is the recognition of the existence of two, in some ways, competing terms, namely *production* and *operations*. The former is the older, and for most people it surely suggests the creation of *goods*, hence the term *manufacture* may be considered synonymous. We do not, however, aim to deal solely with goods creation but also with the provision (or creation) of what for the time being can be called *services*. Thus, for our purposes, operating systems provide goods or give service. (Hence a production system is an operating system but not necessarily vice versa.) Definitions are a matter of preference and of value primarily in communication. Our preference is to define an operating system as a *configuration of resources combined for the provision of goods or services*. Thus operations management is '*the design and planning, operation and control of operating systems*'.

SCOPE, RESPONSIBILITIES AND IMPORTANCE

Operations management is concerned with the acquisition and the use of the (largely) physical resources employed in the provision of goods and services. The scope of the function is adequately indicated by the type of 'headings' normally employed in most texts. These are the traditional problem areas or fields of activity of operations management. They are listed in life-cycle, or chronological, order in Exhibit 2.1, where their relationship with the four activities referred to in our definition is also shown. Key words (i.e. those used later) are in italics. From this it follows that operations management will normally be responsible for the management of inventories, quality, the maintenance and replacement of facilities and scheduling of activities, as well as responsibility for the location, layout, capacity and manning of the system. Managers working in this function will also normally have some involvement in the design or

4

PROBLEM AREAS OR ACTIVITY

Design and Planning
- Involvement in design/specification of product/service.
- Design/specification of process/system.
- *Location* of facilities.
- *Layout* of facilities/resources.
- Determination of *capacity*/capability.
- Design of work or *jobs*.
- Involvement in determination of *remuneration* system, and work standards.

Operation and Control
- *Scheduling* of activities.
- Control and planning of *inventories*.
- Control of *quality*.
- Scheduling and control of *maintenance*.
- *Replacement* of facilities.
- Involvement in *performance measurement*.

Exhibit 2.1 The scope of operations management

specification of the product or services, processes, manning policies and performance measurement.

Whilst this approach to describing operations management gives some indication of scope and responsibilities, it does little to convey the 'flavour' of the function. An adequate treatment of any management subject should offer some guidance on the nature and solution of the problems commonly confronting practitioners. Additionally however the reasons for the existence of such problems, their relative importance, complexity and their interrelationships must be explained. We would hardly be justified in dealing only with problem solutions and must therefore tackle these wider issues.

The operations manager must be familiar with the likely consequences of his decisions in any of the problem areas listed in Exhibit 2.1. He must be able to see his activities and decisions at a policy as well as a functional level. He is directly responsible for the major part of the resources available to the organization. His task is to acquire and utilize labour, materials and equipment—the 'software' and the 'hardware' of the organization. His task is to create and operate an effective system, and he will be required to balance several interdependent, and often competing, objectives. He will be required to satisfy the often conflicting needs of other functions, e.g. the needs and promises of marketing, the requirements of the accountant, and the personnel manager. The operations manager is at the end of the chain of consequence. The consequences of his decisions will be reflected directly in the performance of the organization. Few others within the organization can compensate for his actions, yet often it will seem to him that he is required to compensate for the actions of others!

The task of the operations manager is a complex and an essential one. Numerous examples exist to demonstrate that the success of an organization is as dependent upon the extent of the utilization of its resources as it is upon the nature or quality of those resources. There is little point in employing sophisticated technology and highly skilled labour unless they can be adequately utilized. The operations manager is primarily responsible for ensuring that such a situation exists.

THE SUBJECT OF OPERATIONS MANAGEMENT

The term operations management and the notion of operating systems first began to emerge in management literature in the mid-1960s. One suspects that there were basically two reasons for this development. It was, and still is, argued that much of what used to be discussed under the title of production or manufacturing management was of relevance in non-manufacturing systems, and therefore that there was some justification in considering, under one heading and as one subject, the management of manufacturing *and* non-manufacturing systems. Secondly, it was doubtless the case that at that time production management, whilst recognized as being of considerable importance, had attracted little interest in academic circles. It was convenient therefore to encourage this subject to 'expand' with the prospect of its developing as a more attractive field for study. Such changes began to take place at a time when the manufacturing sector in most developed economies was clearly declining in terms, for example, of total employment, and therefore at a time when increasing attention was being paid to the 'service sector'.

In the past decade the subject of operations management has become established, yet it has to a large extent failed precisely where production management failed. It is still accepted that operating systems for the provision of goods and those for the provision of services do have much in common. It is still argued therefore that operations management, as defined above, can exist as a subject, and that there exists a body of knowledge relevant to this subject. Nevertheless, as we have indicated in Chapter 1, the subject still fails to live up to its apparent potential. Our treatment of the subject is obviously influenced by this perception of developments to date. Clearly the similarities of certain problem areas in different types of system is insufficient proof of the existence of a relevant body of knowledge. Further, the fact that certain techniques have relevance in different systems is an inadequate basis for the development of a subject. Justification for the treatment of operations management as a separate subject depends upon proof of the existence of a substantial and coherent body of knowledge relating to both the theory and practice of 'the design and planning, operation and control of operating systems'. It is clearly insufficient to concentrate upon the management of, say, production systems and to assume the relevance of such a discussion to the management of other systems. Equally it is inadequate to deal with the management of each type of system separately, for this might suggest little fundamental similarity and perhaps the absence of a coherent and relevant body of knowledge. Rather must the fundamental similarities of systems be demonstrated and the differences between them identified. In attempting to make such comparisons the differences within as well as between each of the two categories must be recognized.

We shall examine the nature of operating systems and the nature of the problems facing operations managers, in order to show that different types of operating systems have different characteristics, that is they present different types of problems to operations managers. We hope in this way to show that the nature of the operations manager's job is *in part* influenced by the nature of the system he is required to manage, i.e. what he must do as well as the way in which he must do it is influenced by the problem characteristics of the operating system. For this reason the problem solving procedures, techniques, etc, that he will employ will also be influenced by the nature and characteristics of the system.

Our four principal tasks in this book therefore are as follows:

(1) <u>Task 1</u> We must examine the *nature of operating systems*, in particular
    (a) the *function and objectives* of systems, and
    (b) the *basic types* of system.
(2) <u>Task 2</u> We must examine the nature of the principal *operations management problems*.

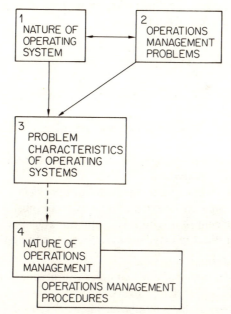

Exhibit 2.2 Concepts for operations management

(3) Task 3 We must look at each type of operating system in the light of these principal problem areas, in order to identify the *problem characteristics of operating systems.*

(4) Task 4 We must consider the *nature of operations management* and the *procedures* and techniques employed in each system.

Exhibit 2.2 illustrates the logic of our approach and provides a framework for the remaining chapters, i.e.:

Section of Book	*Chapters*	*Task (above)*
Section 1	3, 4, 5	1 Nature of operating systems
Section 2	6	2 Operations management problems
Section 3	7, 8, 9	3 Problems characteristic of operating systems
Section 4	10, 11, 12	4 Nature and procedures of operations management

Since so few people read the prefaces or introductions to books it is worth emphasizing that throughout we shall seek to identify the fundamental *concepts* relevant to an understanding of the design and planning operation and control of operating systems. We shall not provide a practitioner's handbook, or an academic treatise, but simply an introduction, and basis for further study. We hope to provide a structure and set of ideas to facilitate understanding of the *nature* of operating systems and operations management. We consider this to be a prerequisite to effective practice in the area.

SUMMARY OF CHAPTER 2

An *operating system* is a configuration of resources combined for the provision of goods or services.

Operations management is the design and planning, operation and control of operating systems.

The management of operating systems is concerned largely with the acquisition and use of *physical resources.*

Operations management will normally hold *responsibility* for the location and layout of facilities, capacity determination, design of jobs, scheduling of activities and the management of inventories, quality, maintenance and replacement.

Our study of operations management requires an examination of:

(1) The *nature of operating systems,*
(2) The nature of the *principal problem areas* of operations management,
(3) The characteristics of each type of system, and
(4) The *nature of operations management* in each type of system and value of available methods, *procedures* and techniques for the solution of management problems.

Section 1

The Nature of Operating Systems

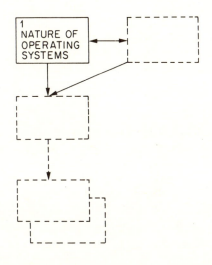

In this section we shall examine the nature of operating systems, identify basic types of systems, and the factors which influence or give rise to the existence of each type of system.

CHAPTER 3

Functions and Objectives of Operating Systems

An operating system has been defined above as a configuration of resources combined for the creation of goods or services. Reference to the act of creation suggests the existence of *productive* systems, indeed operations management could well be defined as the design, planning, operation and control of productive systems. The term 'productive' in turn suggests a basis for the evaluation of the worth of operations management, i.e. how productive or, more generally, how efficient or how effective? If we accept that management in any context seeks to satisfy certain performance criteria then operations management would seem to be concerned with those types of productive work which are geared to the criteria of *efficiency* or *effectiveness*. The definition of an operating system suggests also the process of conversion, i.e. resources into goods or services. In view of the different nature of goods and services we accept that this conversion process might not simply consist of physical conversion, e.g. change of shape or content, but conversion into any output required by the customer, i.e. an output which has some utility for the customer.

Summarizing therefore, an operating system

converts, using *physical resources*,
to create outputs the *function* of which is to satisfy customer wants, i.e.
to provide certain utility for the customer,
whilst satisfying the *objectives* of effective or efficient operation.

These three key aspects of *resource*, *function* and *objectives* will be examined in the following sections.

RESOURCES IN OPERATING SYSTEMS

We have argued, in Chapter 2, that operations management is concerned essentially with the use of physical resources, i.e. their specification, acquisition, deployment, maintenance and replacement. Obviously other types of resource will necessarily exist in operating systems, for example capital will be employed, information will be used and generated, and energy will be required. We might argue that operations managers are generally responsible for more capital than most other managers, that information is the only truly essential resource, and that of course in some situations, e.g. libraries, computer systems, etc, information is a principal product. For our purposes, however, in attempting to understand the nature of operating systems, *we will focus throughout upon physical characteristics.*

We shall concentrate upon physical resources categorized for convenience as follows:

(1) *Materials*, i.e. those physical items consumed or converted by the system, for example raw materials, fuel, indirect materials, etc.
(2) *Machines*, i.e. those physical items utilized by the system, for example tools, vehicles, building, etc.
(3) *Labour*, i.e. those persons who necessarily provide or contribute to the operation of the system, without whom neither machines nor materials are effectively used.

THE FUNCTION OF OPERATING SYSTEMS

In abstract terms the function of an operating system is the satisfaction of customer wants through the provision of outputs of goods or services having some utility.

Outputs might be considered in terms of their tangibility, tangible items perhaps being considered as goods and non-tangible outputs as services. The classification in Exhibit 3.1 is based upon such an approach. The distinction is a useful one, but examination of the examples in Exhibit 3.1 reveals almost as wide a difference in the nature of the items within each list as between both lists. Tangible outputs for example comprise items which themselves might constitute resources for the creation of other such items, e.g. metals into electrical goods, and fabrics into clothing. Furthermore such items will not satisfy customer wants until they are in the 'possession' of the customer. Thus goods may require a service, e.g. advertising, to reach the customer.

Tangible	Intangible
Metals	Medical care
Fabrics	Advice
Food	Entertainment
Electrical goods	Rest
Clothing	Information
Buildings	
Meals	
Electricity	
Heat	

Exhibit 3.1 Operating system outputs

Several points are evident from the above. Firstly it is clear that few organizations will provide either goods *or* services. Many will provide both. Furthermore *within* such organizations almost certainly resources will be utilized for the creation of goods *and* services. If we take the case of a company manufacturing electrical domestic appliances, resources are clearly used for the creation of goods, i.e. appliances, for the satisfaction of customer wants. The same customers may also want, and be provided with, a service, for example a maintenance service for the upkeep and repair of previously supplied appliances. There may also be some form of customer advice provided as a form of service. Within this company there will be assembly lines and machine shops where resources are used for the creation of goods—initially parts and ultimately appliances. Additionally there will be stores, canteens, and maintenance departments, etc, all of which will exist to provide a service. From the point of view of *external* customers, companies ostensibly involved in manufacture, e.g. IBM, General Motors, ITT, etc, devote considerable resources to the provisions of services upon which their manufacturing activities ultimately depend. Equally, organizations ostensibly concerned with the provision of services, e.g. airlines, hotels, etc, devote considerable resources to the provision of goods as a necessary aspect of their overall activities. From an *internal* viewpoint a surprisingly large proportion of all physical resources is often utilized for what might seem to be 'secondary' activities. For example it has been estimated that approximately 50% of the employees of IBM are concerned with non-manufacturing activities.

To categorize organizations as being concerned with the creation of goods or services is clearly an oversimplification. Most are concerned with both, and both aspects are mutually dependent.

In passing, a second point relating to the above is worth noting. For any organization, and particularly ones like IBM or airlines, a large number of 'customers' can be identified. The organization itself has customers, i.e. external customers for its goods and services. Additionally virtually every person and department within that organization will also be a customer for goods or services created within the organization. Assembly departments will be customers for machining departments; design departments for development departments etc. Large organizations can therefore be considered as 'clusters' of operating systems each of which may resemble a larger part of a smaller organization. Thus for the purpose of examining the *nature* of operating systems it may be misleading to compare, for example, an airline and a small retail shop, since the former will in effect contain many of the latter along with many other operating systems. This is the system/sub-system problem which we shall return to in Chapter 4. For the present it is sufficient to note that in any attempt to develop a classification of operating systems it will be easier to compare systems of similar levels of complexity.

Now it is evident that the distinction or classification of operating system outputs as goods or 'service giving' is an inadequate basis for the examination of the function of systems. We have noted the large variety within each of the categories, which would seem to suggest the need for further sub-division.

Such an approach will be adopted below. We will reconsider the function of systems, and must therefore set aside conventional perceptions of what constitutes goods and service sectors in industry, particularly the latter.

Why do operating systems exist, in other words what are their principal functions as perceived by their external customers ? The list of examples given in Exhibit 3.2, whilst by no means comprehensive, is sufficient to indicate the range of functions which might be encountered in operations management. With the possible exception of the coal mine and food canner they might all exist on a similar scale. This list can be used as our 'acid test'. If we are to deal with operations management rather than manufacturing management, then, since these are all examples of systems for the creation of goods or services, we must

---

1 Retail shop (or supermarket)
2 Taxi service
3 Emergency ambulance service
4 Coal mine
5 Tailor ('off the peg' and 'bespoke')
6 Dentist
7 Furniture removal service
8 Fire service
9 Refuse (or 'trash') removal service
10 Petrol (or 'gas') filling station
11 Hospital accident ward
12 Chinese 'take-away' (i.e. take-away or 'eat out' food shop)
13 Launderette (i.e. coin operated laundry)
14 Motel
15 Food canner
16 Builder
17 Broker

Exhibit 3.2 Examples of operating systems

be able to apply our concepts, principles, analyses, etc, to these seventeen examples.

The majority of these examples would *normally* be considered to be service giving systems (e.g. examples 3, 6, 7, 8, 9, 11, 14, 17 and perhaps also 1 and 10). However the diversity of these examples reinforces the point made above, i.e. the need for further sub-division. Developing the notion of function, some further categorization of these seventeen examples is possible.

Manufacture is an obvious category, and would here include the tailor, Chinese take-away, food canner, builder, and perhaps the coal mine. This would seem to be a fairly homogeneous category having one predominant common characteristic, i.e. physical creation. The Chinese take-away may be considered to operate in the manner of a shop; however the predominant characteristic is manufacture, i.e. the manufacture of take-away meals. The system therefore has fundamentally the same function as the tailor, etc.

Transport is an obvious category which would here include the taxi service, emergency ambulance service, furniture removal, and refuse removal. Here again the examples have an obvious similarity in that from the point of view of external customers each exists for the purpose of moving something or somebody from place to place. We may be tempted to think of, for example, the refuse removal as a service. We tend to consider it as part of the service industry, however from our present *functional* viewpoint the principal purpose is to move something, in this case refuse, usually from a house to a disposal site.

Of the remaining examples, the retail shop, petrol station, and broker have an obvious similarity in that they are primarily concerned with the provision or supply of existing items to customers.

This leaves the dentist, fire service, launderette, hospital accident ward and motel—a seemingly heterogeneous group, but one in which each example has at least one important characteristic in common. Their function is to treat their customers whether these be persons, e.g. in need of a repair (hospital accident ward) or a filling (dentist), or items, e.g. in need of extinguishing (fire service) or cleaning (launderette). We shall refer to this class as *service*, despite the fact, as we have seen, that traditionally the term 'service' or the 'service sector' might imply a somewhat broader category than we shall employ.

This rough categorization of operating systems by function therefore provides the following four classes:

(1) *Manufacture* in which the principal common characteristic is that something is physically created, i.e. the output consists of goods which differ physically in form, content, etc, from those materials input to the system. Manufacture therefore requires some physical transformation or a change in *form utility* of resources.

(2) *Transport* in which the principal common characteristic is that someone or something belonging to the customer is moved from place to place, i.e. the location of someone or something is changed. The system utilizes its resources primarily to this end, and such resources will not normally be substantially physically changed. There is no major change in the form of resources, and the system provides primarily for a change in *place utility*.

(3) *Supply* in which the principal common characteristic is that the ownership or possession of goods is changed. Unlike manufacture goods output from the system are physically the same as those input. There is no physical transformation and the system function is primarily one of change in *possession utility* of a resource.

(4) *Service* in which the principal common characteristic is the treatment or accommodation of something or someone. There is primarily a change in *state utility* of a resource. Unlike supply systems the state or condition of physical outputs will differ from inputs by virtue of having been treated in some way.

Exhibit 3.3 summarizes the above.

PRINCIPAL FUNCTION	PRINCIPAL COMMON CHARACTERISTICS	EXAMPLES OF ORGANIZATIONS
Manufacture	Physical creation, i.e. change in form utility of resources	Coal mine Tailor Chinese take-away Food canner Builder
Transport	Change in location, i.e. change in place utility of resources	Taxi service Emergency ambulance service Furniture removal Refuse removal
Supply	Change in ownership or possession, i.e. change in possession utility of resources	Retail shop Petrol station Broker
Service	Treatment of something or someone, i.e. change in state utility of resources	Dentist Fire service Launderette Hospital accident ward Motel

Exhibit 3.3 The function of operating systems

No such categorization by function can be completely watertight, hence some overlap is evident. For example, there is some similarity between transport and service in that transport 'treats' customers to movement. Service may treat customers to such an extent as to physically convert or perhaps create, e.g. a dentist. Further sub-division might eliminate such overlap. However, for our largely descriptive purposes, this categorization will suffice. This descriptive categorization refers to the purpose or function of systems, *not* necessarily their nature. (We shall develop a different approach to categorization in later chapters.) We can now provide alternative definitions for operating systems and operations management, both of which provide a clearer indication of the *scope* of our subject, i.e.:

An *operating system* is a configuration of resources combined for the purposes of manufacture, transport, supply or service.

Operations management is concerned with the design and planning, operation and control of systems for manufacture, transport, supply or service.

We might employ these definitions in describing organizations as a whole (as in Exhibit 3.2), or parts of those organizations. For example, a dentist's surgery has been categorized as a service system, i.e. a system utilizing resources for the treatment of customers. Equally within this system the hygienist operates a service system. A motel provides a service by accommodating people and cars, whilst the restaurant within the motel also provides a service for motel customers. The hospital accident ward provides a service by treating and accommodating patients whilst within that system there may well be manufacturing systems producing medications, etc. We will examine this system/sub-system issue in Chapter 4.

OBJECTIVES OF OPERATIONS MANAGEMENT

Certain organizations must of necessity make a profit, i.e. total revenues generated must exceed total outgoings or costs. In such cases the concept of profit might provide a sufficient basis for establishing objectives for operations management. However some operating systems cannot easily be judged against this objective. For example the hospital, fire service, dentist etc., will not normally be required to maximize profit. In such situations other objectives must be established, whilst even in profit oriented organizations some more immediate objectives for operations management are desirable.

Customer satisfaction

The objective of operating systems has been shown to be the provision of goods or services for the satisfaction of customer wants. In so much as operations management is responsible for the performance of operating systems, this objective of customer satisfaction must also be an objective of operations management. Although at a detailed level the nature of customer wants and the basis of their satisfactions will depend upon the nature, i.e. function of the system, they are similar at a general level. Exhibit 3.4 identifies the main sources of customer satisfaction for each system function. The primary factor is the nature of the goods or services provided. Secondary factors can be considered to relate to costs and timing.

Thus using the classic 'catch phrase', one objective of operating systems and operations management is to provide customer satisfaction by providing the 'right thing (goods or service) at the right price and at the right time'. We shall refer to this as the objective of *customer service*.

Efficiency and effectiveness

Given infinite resources any system, however badly managed, might perform satisfactorily in respect of customer service. Many organizations have gone bankrupt despite having loyal and satisfied customers. The problems, indeed the need for operations management, stems from the fact that operating systems must satisfy multiple objectives. Customer service must be provided simultaneously with the achievement of effective or efficient operation, i.e. effective or efficient utilization of resources. Either inefficient use of resources *or* inadequate customer service is sufficient to give rise to the 'commercial' failure of the operating system.

SYSTEM FUNCTION	SOURCES OF CUSTOMER SATISFACTION	
	Primary factors	Secondary Factors
Manufacture	*Goods* of a given, requested or acceptable specification	Cost, i.e. purchase price or cost of obtaining goods Timing, i.e. delivery delay from order or request to receipt of goods
Transport	*Movement* of a given, requested or acceptable specification	Cost, i.e. cost of movement Timing, i.e. (1) duration or time to move (2) wait, or delay from requesting to its commencement
Supply	*Goods* of a given, requested or acceptable specification	Cost, i.e. purchase price or cost of obtaining goods Timing, i.e. delivery delay from order or request to supply, to receipt of goods
Service	*Treatment* of a given, requested or acceptable specification	Cost, i.e. cost of treatment Timing, i.e. (1) duration or time required for treatment (2) wait, or delay from requesting treatment to its commencement

Exhibit 3.4 Factors in customer satisfaction

Efficiency is conventionally defined in physical terms, e.g. 'the ratio of useful work performed to the total energy expended', or 'the ratio of useful output to input'. Using such definitions efficiency would take a value between zero and unity. It has been pointed out that although this concept might be of relevance in essentially physical activities, for example manufacture, it is inappropriate in respect of organizations as a whole, since as we have noted in many cases the objective will be to output *more* than is input, i.e. the concept of profit or 'value-added'. For this reason the term *effectiveness* might be preferred since it has broader connotations suggesting perhaps the extent or degree of success in the achievement of given ends. Since operations management is concerned essentially with the utilization of physical resources, an objective must be the maximum utilization of such resources, i.e. obtaining maximum effect from such resources or minimizing their loss, underutilization or waste. The extent of the utilization of the potential of resources might be expressed in terms of the proportion of available time used or occupied, space utilization, levels of activity, etc. In each case the objective will be to indicate the extent to which the potential or capacity of such resources is utilized. We shall refer to this as the objective of *resource productivity*.

Operations management is concerned with the provision of *both* satisfactory customer service *and* resource productivity. Other management functions will of course have similar objectives. Those involved in marketing for example will be concerned with customer service and the utilization of the resources at their disposal, e.g. the sales force. Operations managers may be responsible for a large portion of total resources, and for this reason they must attempt to balance their two objectives. Certainly they will be judged against both. One must be balanced against the other since an improvement in one may give rise to a deterioration in some aspect of the other. All of the activities of operations management, i.e. the design and planning, operation and control of the operating system must be tackled with these twin objectives in mind.

This is perhaps an appropriate point at which to remind ourselves that although the twin objectives of operations management are clear, perhaps the manner in which those objectives are pursued, and certainly the emphasis placed on each may be influenced by considerations and decisions beyond the direct influence of the operations manager. We take the view (Chapter 1) that operations management often provides an inadequate contribution to business policy discussions and decisions. Whatever this contribution the requirement of the operations function is one aspect only in the development of a business policy. Other considerations will also be relevant. To some extent therefore operations management will be required to pursue a stipulated policy as effectively as possible. For example, policy in respect of customer service may well be influenced to some considerable degree by broader business policy considerations. Whilst a mail order firm, a luxury store, and a supermarket are all concerned with the function of supply they each have a different approach to the objective of customer service, hence operations management will not be required to achieve the same standards of service in each case. Often standards or objectives in respect of customer service will be influenced by other functions in the organization.

EXAMPLES—RESOURCES, FUNCTION, OBJECTIVES AND TASKS

Some of the examples employed in developing the classification by function earlier in this chapter are re-examined in Exhibits 3.5 to 3.8. In each case the nature of the physical resources employed in the system, and the customers for that system are indicated. The general nature of the twin objectives of operations management is stated together with a brief description of the activities through which such objectives must be satisfied.

| Function: | MANUFACTURE |
| Example: | CHINESE 'TAKE-AWAY' |

OPERATIONS MANAGEMENT

Objectives   Activities to achieve objectives

Design/ Planning
- Location of shop
- Facilities layout
- Specification of menu
- Selection of ingredients
- Methods of preparation
- Packaging etc.

Operation/ Control
- Determination of opening hours
- Staffing levels
- Amount of equipment and facilities used
- Stock Control
- Maintenance
- Replacement of equipment
- Control of quality

Resource Productivity

e.g. relating to:
Staff Utilization
Equipment utilization
Stock of materials
Material wastages

AND

Customer Service

e.g. relating to:
Menu choice
Cost
Quality of food
Waiting time

Physical Resources:
Cooking equipment
Serving equipment
Ingredients
Indirect materials (e.g. wrappings etc)
Cooking staff
Serving staff
Shop facilities – counter, cash register, storage facilities, furniture, parking space etc.

Customers
Individual members of public

Exhibit 3.5 Chinese take-away example

OPERATIONS MANAGEMENT

| Function: | TRANSPORT |
| Example: | AMBULANCE SERVICE |

Objectives

Activities to achieve objectives

Physical Resources:
Ambulances and auxiliary vehicles
Drivers and attendants
Maintenance facilities and spare parts
Garaging facilities
Communications equipment
Medical equipment
Control staff
Maintenance staff

Resource Productivity
e.g. relating to:
Staff utilization
Equipment utilization

Design/ Planning
Design of equipment
Location of vehicles
Design of communications system
Determination of staffing levels
Number of vehicles
Manning of vehicles
Work shifts
Routing of vehicles

AND

Customers
Individual members of public

Customer Service
e.g. relating to:
Waiting time
Speed of Service

Operation/ Control
Stock control
Secretarial services
Maintenance and replacement of equipment

Exhibit 3.6 Emergency ambulance service example

OPERATIONS MANAGEMENT

Function:	SUPPLY
Example:	RETAIL SHOP

Objectives

Activities to achieve objectives

Physical Resources:

Goods in stock
Shop facilities i.e.
she ving, counter,
cash register, furniture,
spares, shop front
displays, parking space
etc.
Counter staff
Indirect materials e.g.
wrapping, boxes, bags etc.
Auxil ary staff

Resource Productivity

e.g. relating to:
Stock turnover
Staff utilization
Stock losses
Average stock levels

AND

Design/ Planning

Location of shop
Facilities layout
Determination of stock
range etc.
Stock levels, brands etc.
Opening hours

Operation/ Control

Stock control
Security service
Maintenance and replacement
of equipment

Customers

Individual members of
public

Customer service

e.g. relating to:
Choice of goods
Costs
Waiting time
Service time

Exhibit 3.7 Retail shop example

Function: SERVICE
Example: MOTEL

OPERATIONS MANAGEMENT

Objectives | Activities to achieve objectives

Physical Resources:
Bedrooms
Amenities and common facilities e.g. dining room, lounges etc.
Service facilities and equipment
Car park
Reception staff
Domestic and auxiliary staff

Customers
Members of the public, individually or in groups

Objectives

Resource Productivity
e.g. relating to:
Room occupancy
Staff utilization
Stock levels
Utilization of other facilities

AND

Customer service
e.g. relating to:
Cost
Room availability and choice
Facilities available

Activities to achieve objectives

Design/Planning
Location of facilities
Layout
Specification of range of facilities
Number of bedrooms, size etc.
Number of staff
Capacity of other facilities
Operating hours and periods

Operation/Control
Security
Service, maintenance and replacement of facilities
Stock control

Exhibit 3.8 Motel example

SUMMARY OF CHAPTER 3

An operating system converts *physical resources* to create outputs the *function* of which is to satisfy customer wants, whilst satisfying the *objective* of effective or efficient operation.

The principal physical *resources* employed are materials, machines and labour.

The *functions* of operating systems can be categorized as manufacture, transport, supply and service.

An *operating system* is therefore a configuration of resources combined for the purposes of manufacture, transport, supply or service.

The twin and perhaps to some extent competing *objectives* of operations management are customer satisfaction and resource productivity.

CHAPTER 4

The Structure of Operating Systems

We have chosen to deal with the management of systems for manufacture, supply, transport and service. Such a categorization is of value for descriptive purposes, but it is of little or no value as a basis for studying operations management. It has been convenient to adopt these terms in identifying the *scope* of operations management. They *each* suggest something in that for example the class of systems referred to as manufacture have something in common and are distinct from those concerned with supply etc. This categorization tells us something about the function of systems but little about the *nature* of such systems. We need to identify the similarities between systems. We should categorize them, but must do so in a way that is meaningful from the point of view of operations management. We must examine the nature and the similarities of the problems facing operations management. This, in turn, requires an understanding of the nature and characteristics of systems. In this chapter, therefore, we will look more closely at the systems in our four categories of operating system—manufacture, supply, transport and service—concentrating on their *structure* or configuration. We shall look particularly at those aspects of operating systems which might be determined or influenced by factors normally beyond the immediate control of the operations manager, since such factors will influence the nature of the problems facing operations management.

Inevitably we must adopt some form of 'systems' approach to the study of operating systems. In the following section we will present some key concepts of 'systems theory' which will be of relevance to our examination of operating systems.

SYSTEMS CONCEPTS

Definition

We shall define a system as *a configuration of entities and the relationships amongst these entities*. The decision to refer to entities or 'things' in systems is significant since we shall concentrate on those physical entities previously defined as resources.

A system can generally be identified where a set of entities have some functional or purposeful interdependence, i.e. where together they constitute a

26

meaningful whole. Virtually everything therefore can be seen as a system or a part of a system. Virtually all systems therefore exist within others. They are all sub-systems of larger systems. Systems form hierarchies, with each system encompassing a larger number of entities and their interrelationships than those systems lower in the same hierarchy. It follows therefore that irrespective of the system considered some external relationships will exist.

The simplest system structure or model utilizes three types of entities, i.e. inputs, processes and outputs, having interrelationships of the form shown in Exhibit 4.1. The amount of detail used in the representation of any system is a

Process

Input (s)   Output (s)

Exhibit 4.1 Simple system model

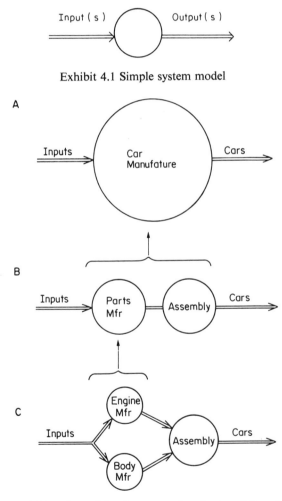

Exhibit 4.2 Models of a system (employing different
amounts of detail)

matter of choice. For example the manufacture of a car may be considered as a single process as in Exhibit 4.2A. Employing more detail two processes might be identified, i.e. parts manufactured and assembly (Exhibit 4.2B), or three processes, i.e. engine manufacture, body manufacture and vehicle assembly.

The selection of the system boundaries is also a matter of choice. In fact the choice of what to include, and the amount of detail employed in representing it will normally be influenced by the objectives or purposes of the exercise. The simple input(s)/process/output(s) system of Exhibit 4.1 could represent a company, a department in that company, an operation within that department etc.

Control

Control is an essential feature of most systems. In most artificial (i.e. man-made) and natural systems some degree of control is normally evident. Classically there exists two basic configurations of control, i.e. *open* or *closed* systems.

An *open system* exists where the outputs have no direct influence over earlier parts of the system. In other words an open system does not react to its own performance. Its past actions have no influence over current or future actions. There is no 'feedback' of information on its output(s) for the control of its input(s). A clock is an example of a simple open system.

A *closed system* is directly influenced by its own past behaviour. Its own output(s) are monitored or observed in order that some purposeful control might be exercised over its input(s). The operation of the system is dependent upon direct feedback of information. The normal domestic central heating system is an example of such control, since the thermostat monitors room temperature and controls the boiler to maintain a given room temperature. This simple concept of feedback control is illustrated in Exhibit 4.3.

Some control may derive from the monitoring of things outside the defined system. Some central heating systems are controlled by a thermostat placed

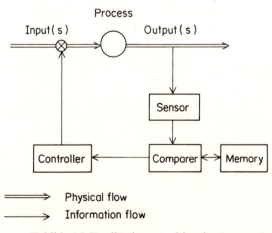

Exhibit 4.3 Feedback control in a system

outside the house monitoring outside temperature. In such cases there is some control but the system is an open system. To some extent therefore whether a system is considered open or closed is dependent upon the choice of the system boundary. Exhibit 4.4 illustrates this. With the boundaries employed in case A the system is closed but with the boundaries employed in case B neither system is closed.

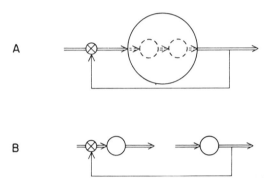

Exhibit 4.4 Closed and open systems

In practice the controls on and within systems will be complex and in many cases entities and their interrelationships may be subject to both internal and external controls. Thus closed systems may also be subject to control from outside defined system boundaries. Since the outputs of some systems will be the input of others these outputs may also be controlled. In fact because of the hierarchical nature of systems, virtually all systems will be influenced by external factors. Exhibit 4.5 illustrates.

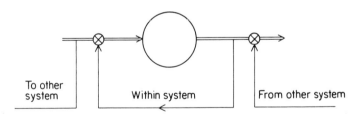

Exhibit 4.5 A system and its relationship with external systems

SYSTEM STRUCTURE

Our objective in this chapter is to achieve a better understanding of the nature of operating systems than that provided through categorization by function. Such an understanding is a prerequisite for the identification of the problems of managing operating systems. We are interested in identifying those similarities and differences between systems which are of relevance to operations management. We propose to do this through the examination of *systems structure.*

System level

We intend to examine the relationship between operating systems structures and operations management problems. In other words we shall identify the influence of structure on the nature of operations management. Such an approach should enable us to describe the nature of operations management in different situations. If we choose to discuss system structure at a relatively abstract, that is high, level, then we shall in turn be examining the nature of operations management at this same level, e.g. the management of operations for the whole organization. Thus we might expect to learn something of the general nature or character of the management of the organization. If we consider systems at a lower level, then we shall be examining operations management at a similar level, e.g. the nature of the management of processes or small departments.

It is a matter of convenience at which level we choose to work. Since we have tended to consider the operations management job primarily in terms of relatively small systems, e.g. departments, small organizations, etc, rather than major complex organization, we will where possible consider system structure at a similar level.

Structure

We have seen that the simplest descriptions of system structures involve 'inputs', 'processes' and 'outputs'. In the context of operating systems, these three terms might be interpreted as follows:

(1) Inputs include: materials
                               machines
                               labour
                               capital
                               information
(2) Outputs include: goods
                               service
                               byproducts
                               waste
(3) Processes involve: conversion in respect of location,
                               physical characteristics,
                               ownership or possession, or state

We have defined operations management as being concerned primarily with physical resources hence for our present purposes it is appropriate to consider inputs to be materials, labour and machines. The interpretation above suggests that the 'process' be considered as a conversion operation of one of the four *functional* forms identified earlier, i.e. manufacture, supply, location or service.

Thus ignoring byproducts, waste, etc, we have the generalized operating system structure shown in Exhibit 4.6.

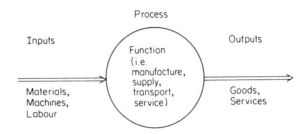

Exhibit 4.6 A simple operating system model

We have noted that a simple structure of this type might represent any operating system and at any level. Thus as a means of examining the nature of operations management this approach is trivial. We must examine system structure in more, i.e. sufficient, detail. Such detail might be represented by the use of models incorporating sets of input, processes, outputs (see Exhibit 4.2). This approach will be used, however, as we have now adopted a rather narrower interpretation of process (i.e. function), and since we are to be concerned primarily with physical resources it is necessary to introduce an additional symbol to represent physical storage of inputs or outputs. Since we are dealing with 'purposeful' systems we can assume that customers exist and can be identified for all operating systems. Such customers or beneficiaries might be subsequent processes, individuals or organizations external or internal to the organization, etc. We shall indicate the 'location' of such customers for each system, thus the notation to be employed is summarized in Exhibit 4.7.

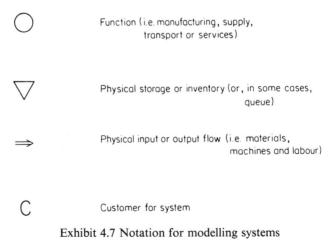

Exhibit 4.7 Notation for modelling systems

Manufacturing Systems

We can identify four simple structures for *manufacturing* systems as shown in Exhibit 4.8. Structure A depicts a 'make to stock, from stock' situation, i.e. all input resources are stocked and the customer is served from a stock of finished

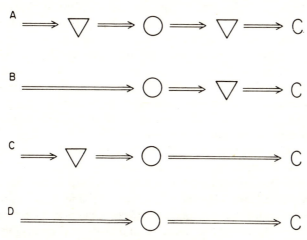

Exhibit 4.8 Structures for manufacturing systems and supply systems

goods. Structure B depicts a 'make to stock, order to make' situation, i.e. no input resource stocks are held, but goods are produced to stock. Structure C depicts a 'make to order, from stock' situation, i.e. all input resources are stocked but goods are made only against and *on* receipt of customers orders. Finally structure D depicts a 'make to order, order to make' situation, i.e. no input resource stocks are held, and all goods are made only against and on receipt of customers orders. Each structure shows how a system will provide for future output. Structure D for example indicates that in order to provide the next output for a customer resources must first be acquired, whereas in C the next customer order will be satisfied using already existing resources. (Examples of each structure are given at the end of the chapter.)

Supply Systems

Now considering *supply* systems in a similar manner, we can again recognize the validity of the four simple structures shown in Exhibit 4.8. Both structures A and B require function in anticipation of order, i.e. structure A depicts supply to 'stock from stock' and structure B depicts 'supply to stock, order to provide'. Neither case is common in supply operations, but both can exist, as will be shown below. More commonly structures C and D will exist. Structure C depicts 'supply to order from stock', whilst structure D depicts 'supply to order, from order'.

Transport Systems and Service

A slightly different situation applies in respect of both *transport* and *service*. All structures which require function in anticipation or in advance of receipt of customer order, are infeasible, since in the case of both transport and service no physical output stock is possible. Consider *transport*. A taxi service cannot satisfy a customer's relocation or movement requirements in advance of receiving the customer's order. Similarly the ambulance, refuse or furniture removal services cannot build up a stock of outputs to satisfy future customer demands. Nor can a 'bus service perform its function of transporting individuals before those individual customers arrive. The 'bus can, and often does, move from stop to stop along its route even though no customers have arrived. In doing so, however, it has not performed its function of changing the location of its customers. In fact it has simply remained as an unutilized stocked resource, in need of customers. Nor can *service* systems such as the fire service, launderette, hospital and motel build up a stock of outputs to satisfy future customer orders.

One further important structural difference is evident in the case of transport and service systems. Since the function of transport and service is to 'treat' the customers (whether a thing or a person) the customer is a resource input to the system, i.e. the *beneficiary of the function is or provides a physical resource input to the function*. Thus transport and service systems are dependent upon customers not only taking their output and in some cases specifying what that output shall be, but also for the supply of some physical input to the function without which the function would not be achieved. For example, in transport, the taxi, ambulance and bus services move the customer or something, the supply of which is controlled by the customer, e.g. a piece of luggage. In service systems, for example the hospital and motel, the customer in person is treated, whilst launderettes and fire services treat items which might themselves be considered as customers (e.g. burning houses) or whose supply is controlled by the customer.

In other words, unlike manufacture and supply, transport and service systems are activated or 'triggered' by an input or supply. The customer exerts some 'push' on the system. In manufacture and supply the customer acts directly upon output—he 'pulls' the system, in that he pulls goods out of the system whether direct from the function (structures C and D) or from output stock (structures A and B). In transport and service the customer pushes the system—he acts directly on input. In such systems therefore some part of the resource inputs are not *directly* under the control of operations management. In these 'push' systems the customer controls an input channel, and we must therefore distinguish such input from that controlled by operations management.

All systems could perhaps be considered as types A, B, C or D; however the use of this push/pull distinction will help us examine the nature and characteristics of operating system.

Somewhat different structures are therefore required to represent transport and service systems. Three structures would seem to exist, as illustrated in Exhibit

4.9. Structure E depicts a function from stock, direct from customer situation, i.e. input resources are stocked excepting in the case of customer inputs where no *queueing* exists. Structure F depicts 'function from source, with customer queue'

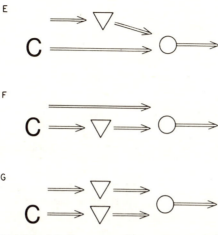

Exhibit 4.9 Structures for transport and service systems

situation, i.e. no input resources are stocked although customer inputs accumulate in a queue. Structure G depicts a 'function from stock with customer queue' situation in which all input resources are stocked and allowed to queue. For simplicity and clarity we shall, when referring to customer push situations, refer throughout to customer queues. Such queues are physical stocks in the customer input channel, although they cannot be utilized by operations management to the same extent as other resource stocks for they are usually beyond their direct control. Queues comprise those customers who have 'arrived' at the system and await service or transport. They are the customers who at any one time have asked to be 'treated' by the system. The queue therefore represents known and committed future demand.

The Seven Basic Structures

The seven basic structures for operating systems are listed in Exhibit 4.10. The mnemonic code introduced in the Exhibit provides a simple means for systems description. It will be used hereafter when identifying system structures in preference to A, B, C, D, E, F and G*.

Throughout the above discussion we have considered the customer and his requirements as given, that is beyond the direct control or influence of operations management. The nature of the customer's behaviour is a major influence on systems structure. For example we have indicated that one fundamental

---

* These mnemonics will be used throughout the remainder of the book. Readers may wish to copy them and the diagrams from Exhibit 4.10 to a separate piece of paper or 'bookmark' for ease of reference.

BASIC STRUCTURE	DESCRIPTION	MNEMONIC*
A $\Rightarrow \triangledown \Rightarrow \bigcirc \Rightarrow \triangledown \Rightarrow C$	Function from stock, to stock, to customer	SOS
B $\Longrightarrow \bigcirc \Rightarrow \triangledown \Rightarrow C$	Function from source, to stock, to customer	DOS
C $\Rightarrow \triangledown \Rightarrow \bigcirc \Longrightarrow C$	Function from stock direct to customer	SOD
D $\Longrightarrow \bigcirc \Longrightarrow C$	Function from source direct to customer	DOD
E $C \overset{\Rightarrow \triangledown \Rightarrow}{=\!=\!=} \bigcirc \Longrightarrow$	Function from stock, and from customer	SCO
F $C \overset{=\!=\!=}{\Rightarrow \triangledown \Rightarrow} \bigcirc \longrightarrow$	Function from source, and from customer queue	DQO
G $C \overset{\Rightarrow \triangledown \Rightarrow}{\underset{\Rightarrow \triangledown \Rightarrow}{}} \bigcirc \longrightarrow$	Function from stock, and from customer queue	SQO

*Mnemonic code

O = Operation —O in middle represents a customer 'pull' system
O at end represents a customer 'push' system
S = Physical stock—S at beginning represents an input stock
S at end represents an output stock
D = Direct —D at beginning represents direct input
D at end represents direct output
C/Q= Customer —C in middle represents direct entry
Q in middle represents queuing

Exhibit 4.10 Seven basic structures for operating systems

feature of transport and service systems is their dependence upon inputs controlled by the customer. In practice an organization may well have some scope for influencing the customer, and thus the pull or push on the operating system. Such influence may derive from advertising and marketing activities, pricing, product policies, etc. For example although the operators of car ferries between the UK and Europe cannot determine customer behaviour since they cannot change the public's wish to travel in July and August and at weekends, there is perhaps some scope for influencing customer behaviour through fare structures. However fares and other concessions and inducements will not usually be under the *direct* control of operations management. An organization may have *some* influence over the structure of the operating system. Such influence may permit the adoption of particular structures as a matter of preference or policy. It may be able to create a situation in which customers for transport or service will be willing to queue, and thus may be able to create a structure SQO situation with enhanced productivity whereas otherwise only a structure SCO system would have given a level of service acceptable to the customer. However the operations manager will not normally have such *direct* influence or control. Furthermore for our present purpose, i.e. that of identifying basic structures of operating systems and their relation with operations management, we can consider the customer as a form of given constraint or influence, at least as far as the operations manager is concerned, and can therefore utilize the seven basic structures for operating systems as the basis for our study of operations management.

To some extent we can now ignore the function of systems since in our study of the nature of operations management we need only be concerned with the structure of systems. The *basic* nature of the problems facing operations managers will, we believe, depend very much upon the structure and far less on the function of the system.

These are of course simplified structures, particularly since in considering those input resources which are under the control of operations management (i.e. all but customers) we have recognized only a single state for each structure. In fact, considering only physical input resources we have already identified three broad categories (materials, machines and labour). In practice some of these resources may be stocked whilst others are not. For example in manufacture even when materials are not stocked, machines and labour often are. Hence in practice, the number of possible structures for manufacturing systems is in excess of the four shown in Exhibit 4.10. We shall utilize these four basic structures in our analysis for the present although strictly such structures apply only for *single-channel* systems, i.e. those with single controlled input channels. We do recognize multiple channels in structures SCO, DQO and SQO in order to indicate that some input is controlled by the customer. We nevertheless make the same simplifying assumption, by showing only a single channel for the other resource inputs. In fact our basic structures will accurately represent multiple channel systems where each type of input resource is treated similarly in respect of the provision or absence of stock. In other cases two or more basic structures will be required to represent systems as in Exhibit 4.11 which illustrates one multi-channel arrangement.

In systems with structure SOS, DOS, SOD or DOD materials or goods flow through the system, whilst in systems with structure SCO, DQO or SQO the flow comprises customers or customer-provided items. In managing manufacture and supply systems we often take the non-consumables for granted and deal largely with consumable items. In transport and service systems we tend to think mainly of non-consumable items (e.g. vehicles, equipment, etc) and ignore consumed items (e.g. petrol). If we take this approach, we can quite easily view all systems as having a single input channel, i.e. that of the dominant or most important input for our purposes. We will adopt this approach in identifying examples below and in considering issues during subsequent chapters. In many cases a single channel model will accurately represent the system, whilst in other cases we must take the approach illustrated in Exhibit 4.11. This latter approach will be developed in Chapter 14.

36

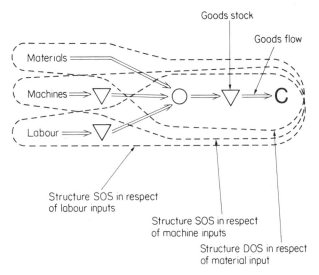

Exhibit 4.11 A multi-input channel system

We should also recognize that some systems have two or more outputs, some of which are stocked, i.e. such systems have two types of output channel. This can apply only in the case of structures SOS, DOS, SOD and DOD and will in fact apply only in manufacturing systems. Thus in certain cases the resources available to the system will be utilized in the manufacture of a range of goods, some of which may be 'made to stock' whilst others are 'made to order'. We shall explicitly consider only the normal single output or identical output channel cases. An approach similar to that shown in Exhibit 4.11 might be employed in considering other cases, or alternatively in such cases we can consider only the dominant or most important output.

BASIC STRUCTURES—EXAMPLES

Structure SOS $\Rightarrow \triangledown \Rightarrow \bigcirc \Rightarrow \triangledown \Rightarrow C$

Such a structure might represent batch *manufacture* of a range of standard products or mass manufacture of products where demand fluctuates. An off-the-peg tailor will often engage in batch manufacture, and of course most engineering production is based on batch working, in both cases an output stock of products being required since during the period of manufacture the output rate will exceed demand rate. In mass manufacturing as, for example, in a coal mine, some food production, vehicle production, etc, an output stock will normally be required since output rate will be reasonably constant whereas demand may well fluctuate. The alternative approach (structure SOD) would involve frequent change in the production rate which, with the type of resources involved, would often be prohibitively costly.

In *supply*, structure SOS might depict a system whereby a supplier reserves (and therefore commits) stocked items for a customer who then 'calls off' small quantities from this, in effect output, stock over a period of time. In such a situation the supplier is in fact holding the customers input stocks, an arrangement which is sometimes employed in small retail outlets when valued customers regularly purchase particular items which must be acquired in large numbers, yet are not requested by other customers. A similar arrangement can often be found in certain suppliers serving the engineering industry, e.g. steel stockists.

Structure DOS $\Rightarrow \bigcirc \Rightarrow \triangledown \Rightarrow \mathsf{C}$

This basic structure might represent batch *manufacture* of products using resources which are only occasionally (or seasonally) available, e.g. a form of opportunistic manufacture, or alternatively the use of input resources which are susceptible to deterioration. Such a situation might more frequently apply in respect of material inputs and an example in such a case might be certain types of food canning and of course vintage wine making. (We should perhaps ignore the case of companies which having been set up to make certain items are forced to close after a short production life!)

In the case of *supply* systems this structure might represent the situation in which the supplier directly orders items specifically for a customer and holds them in stock for him, the customer then 'calling off' from this stock in small quantities. Again the structure is more likely to apply in respect of material inputs.

Structure SOD $\Rightarrow \triangledown \Rightarrow \bigcirc \Rightarrow \mathsf{C}$

Manufacture from stock direct to customer suggests the production of 'one-off' items or special versions of the given item type against individual customer orders. Alternatively manufacture as required of a limited range of items which cannot be stored would also correspond to structure SOD. Examples (from Chapter 3) might include a bespoke tailor. A Chinese take-away will often have structure SOD since meals will normally be prepared against each customer order albeit from precooked ingredients. (In contrast, take-away or 'eat-out' food shops with a simple and limited menu, e.g. hamburger, 'houses' such as MacDonalds in the USA, or fish and chip shops in the UK will often operate with structure SOS.) For *supply* functions structure SOD indicates a system whereby the supplier provides items from stock on demand. Most retail and wholesale supply operations are of this form. The retail shop and petrol station would normally have structure SOD.

Structure DOD $\Rightarrow \bigcirc \Rightarrow C$

This structure suggests the *manufacture* of 'pure one-offs', i.e. items requiring unique resource inputs or items manufactured on a very infrequent basis. Certain major civil engineering work may be of this form. The establishment of facilities and the building of oil rigs for North Sea oil exploration and production in the early 1970s may be seen as an example of this structure. Frequently the system will apply only in respect of consumable inputs, i.e. materials and to a lesser extent labour. Some house building is undertaken in this manner.

Structure DOD for *supply* suggests an arrangement whereby a supplier (e.g. an agent, 'middleman' or 'broker') acquires direct for his customer. In such cases the structure applies strictly in respect of single-channel, i.e. material inputs. Examples (in respect of material inputs) might include certain building supplies and the purchase and supply of some commodities. Some real estate agents used to work in this fashion although the majority act for house sellers and in effect operate with structure SOD.

Structure SCO $C \rightrightarrows \bigtriangledown \Rrightarrow \bigcirc \Rightarrow$

Transport with structure SCO will exist where customer queueing cannot be tolerated, is impractical, or impossible. Taxis and emergency ambulance services normally correspond to this structure.

In the case of *service* a similar situation applies, i.e. where customer queueing does not take place for one reason or another. A hospital accident ward, fire service, and motel will normally have this structure. Notice that systems which service customers who are unwilling to queue must fall into this category. It is for this, and of course cost reasons, that many petrol stations now provide self-service. Drive in or drive through banks, drug stores and other convenience (i.e. little waiting) shopping is provided for much the same reasons. A launderette will often work as structure SCO since there will be insufficient customers to necessitate their queueing.

Structure DQO $C \rightrightarrows \bigtriangledown \rightrightarrows \bigcirc \Rightarrow$

This structure depicts the situation in which customers queue and await the function which is not performed until all resources are acquired. Such a situation would apply in respect of both *transport* and *service* where it was necessary to satisfy entirely new or unexpected demands, needs or circumstances. The chartering of aircraft for groups, concert promotion, etc, might reflect structure DQO. On a very large scale the operation of placing a man on the moon was an example of a structure DQO transport system. On a smaller scale certain major engineering work involving transport or movement of major structures might also be an example of structure DQO particularly in respect of labour and machine inputs.

Structure SQO

$$\Rightarrow \bigtriangledown \Rightarrow \bigcirc \Rightarrow$$
$$C \Rightarrow \bigtriangledown \Rightarrow$$

Numerous examples exist both in respect of *transport* and *service*. Transport systems such as refuse removal, furniture removal, etc, are all normally examples of structure SQO; indeed most transport particularly public transport systems correspond to this structure. Similarly most dentists will have this structure, i.e. where customers queue to be 'treated'.

Exhibit 4.12 summarizes by indicating the *normal*, i.e. usual, structures for the seventeen systems identified in Chapter 3. We must emphasize the word *normal*. Since the structure of systems may change we should strictly consider their structure at a point in *time*. It is convenient for our present purposes of illustration however to consider the normal (or usual, or intended) structure of systems. We shall return to the question of structure change, its causes and effects in Chapter 5.

HIERARCHIES OF SYSTEMS

The hierarchical nature of systems should again be noted. This suggests that we might identify examples of these seven basic system structures within larger systems. Considering firstly functions, supply, service and transport systems are often found within manufacturing systems, for example stores, repair and internal transport respectively. Supply and transport systems are found in service systems, e.g. stores and lifts in hospitals and supermarkets. In fact examples of each of the four functions may be identified within larger systems. Similarly, therefore, should we choose to describe any system in more detail, we might identify sub-systems with one or more of the seven basic structures. For example a launderette as a simple system has been considered (above) as an example of structure SCO. Defining the system in more detail however we might identify two sub-systems involving firstly the customer's use of a washing-machine, and secondly the use of a drying machine. Both sub-systems might be examples of structure SCO. Exhibit 4.13 illustrates. A fish and chip shop or hamburger house like MacDonalds might at a simple level be described as structure SOS or alternatively in more detail as two sub-systems.

Clearly, given examination at a sufficient level of detail, any system of one of

40

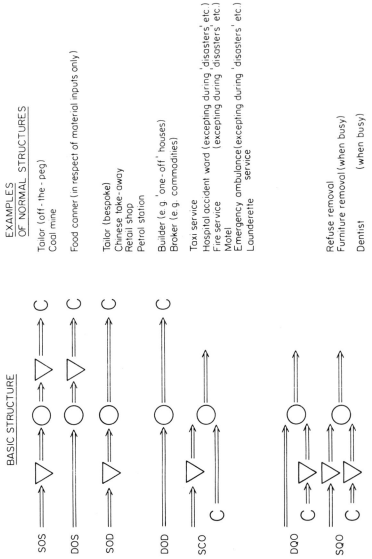

Exhibit 4.12 Examples of basic operating system structures

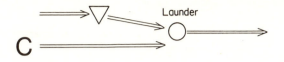

A. A launderette as a 'simple' system

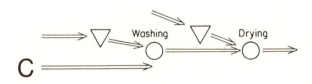

B. A launderette as a 'complex' of basic systems

Exhibit 4.13 Systems within systems

the seven basic structures might be seen to consist of sub-systems themselves of one of these seven structures arranged in series and/or in parallel. We shall return to the examination of such 'complex' systems in Section 4. However the contents of the following chapters, which examine the nature of systems, in particular the operations management problems associated with each of the seven basic types of systems, apply irrespective of the level of detail employed in identifying systems. Thus, for example, the factors influencing the existence of a certain system structure will apply whether we consider the system at the level of the firm or some small part of that firm. Equally the general nature of operations management problems associated with each system structure apply irrespective of the level of the system. Putting it another way, in relation to Exhibit 4.13, when we have identified the factors which give rise to existence of a system with structure SCO, the operations management problems which are features of structure systems, and the procedures which might be relevant in tackling such problems, this information will be of relevance in considering the design or planning, and operation or control of a launderette as well as of the sub-systems of the launderette.

Similarly familiarity with the nature of systems of basic structure SOD, will be of relevance in considering the management of a retail operation as well as in the management of a small storeroom in a manufacturing operating system. The level of systems, i.e. the amount of detail represented by a basic structure, does not affect the relevance of the concepts and issues discussed in the remaining chapters in this section, and those in Section 2.

SUMMARY OF CHAPTER 4

Identification of the nature of operating *system structures* is a prerequisite to an examination of the nature of operations management.

Seven basic operating system structures can be identified, i.e.:

SOS   Function from stock, to stock, to customer.
DOS   Function from source, to stock, to customer.
SOD   Function from stock direct to customer.
DOD   Function from source direct to customer.
SCO   Function from stock and from customer.
DQO   Function from source, from customer queue.
SQO   Function from stock, from customer queue.

The distinguishing features of structures are:

(1) The existence and location of *inventories*, and
(2) The influence of the *customer*, in particular whether the customer pulls the system by acting on demand (as in manufacture and supply systems) or pushes the system by acting on supply (as on transport and service system).

These basic structures are of relevance in describing operating systems at any level of detail, i.e. at any hierarchical level.

CHAPTER 5

Factors Influencing System Structure

We have identified seven basic structures for operating systems. Chapter 6 will examine the relationships of system structure and the problems encountered in managing systems, and thereafter we shall consider in more detail the problems of operations management for given structures. We shall tend to assume the existence of a structure and look at the management of that system. It is appropriate therefore that we consider the *reasons* for the existence of certain structures in this chapter. We must seek to explain why these seven basic structures exist. What factors permit and/or give rise to their existence? To what extent are structures determined by factors beyond the control of operations managers and, conversely, are operations managers able to influence or choose structures?

There are certain prerequisites for the existence of operating system structures. Certain factors will permit, and in exceptional conditions cause, one or more of the structures to exist. Such prerequisite or enabling factors are essentially of an *external* nature and are largely beyond the direct control of operations managers.

To some extent however operations managers can, through their decisions and activities, influence system structure. Some degree of choice may exist. Structures may therefore be manipulated or changed. There are therefore *internally related* factors of relevance to system structure.

EXTERNAL FACTORS

Demand limitations

In certain structures the function relates to the subsequent functions—which will often be the customer—via a stock of items (e.g. structures SOS and DOS). Similarly the relationship of the function to prior functions may be through physical stocks (e.g. structures SOS, SOD, SCO and SQO). In other cases the relationships are direct. The nature of this relationship may be influenced by, not the level of demand or the variability of demand level, but the *predictability of the nature of demand*. In other words, the provision of physical stocks requires

43

some knowledge or predictability of *what* will be required at subsequent stages in the system. Structure SOS, for example, affords function in anticipation of demand, i.e. 'from stock, to stock, to customer', hence the nature of customer demand, i.e. *what* will be required by the customer must be known. Additionally, since input resources are stocked, it is essential that the nature of the function be known. Predictability of function is of course related to predictability of the nature of customer demand. Predictability of the nature of demand is an essential *prerequisite* of structure SOS. The same prerequisite applies for structure DOS even though in this case input resources are not stocked.

A prerequisite for the existence of input stocks is predictability of function. However, since in many systems input resources might be employed to provide a variety of outputs, predictability of function may not depend upon full knowledge of the nature of customer demand. For customer push structures, exactly the same considerations apply. In structures SCO and SQO, some degree of predictability of the nature of demand is a requirement. Again since system input resources may be utilized to provide a variety of output, some degree of uncertainty of customer demand might exist. Unpredictable customer demands necessitate structures DOD or DQO.

Clearly the feasibility of the provision of stocks is dependent upon the predictability of the nature and therefore the needs of subsequent functions. For our purpose this can be expressed in terms of the predictability of the nature of the demands of the customer in the system.

Exhibit 5.1 summarizes. Notice that such predictability is only an enabling factor, and that the predictability of demand influences structure *feasibility*. It

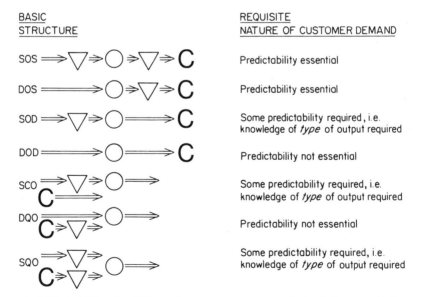

BASIC STRUCTURE	REQUISITE NATURE OF CUSTOMER DEMAND
SOS	Predictability essential
DOS	Predictability essential
SOD	Some predictability required, i.e. knowledge of *type* of output required
DOD	Predictability not essential
SCO	Some predictability required, i.e. knowledge of *type* of output required
DQO	Predictability not essential
SQO	Some predictability required, i.e. knowledge of *type* of output required

Exhibit 5.1 The feasibility of system structure and the influence of the nature of customer demand

does not follow that the existence of predictable customer demand will necessarily give rise to the existence of a certain structure. For example, the nature of the demand for the output of a power station is known, yet output is not stocked. This condition is, however, *one* prerequisite, the *external* one, without which certain structures will not in reality exist. Only in exceptional conditions will this external factor cause the existence of a system structure. Notice, furthermore, that the above discussion relates only to the existence of stocks directly controlled by operations management. The existence of queues of customers (structures DQO and SQO) will be influenced mainly by the internal factors discussed below.

Input limitations

We have considered only one, albeit the most important, aspect of the external environment and its influence on the operating system. The nature of the operating system and the management of that system may be influenced by other external factors, indeed by all the uncertainties facing the system. For example, the uncertainty or predictability of inputs to the system might influence system structure. Two aspects of the predictability of resource input could be of relevance here, i.e. uncertainty of the *nature* of input and uncertainty of input *flow*. Uncertainty of the nature of inputs is unlikely, but possible. Uncertainty about the nature of one input resource may prevent the others being stocked. When, for example, labour and machines are provided in order to work upon, utilize and convert material resources, the nature of the labour and machinery required may be influenced by the nature of the materials to be employed. In such a case, therefore, uncertainty of the nature of one resource input will in effect render certain structures infeasible. Uncertainty of inputs limits system structure *only* in this way.

It is more likely that supply uncertainties will concern input flows, e.g. input quantities such as batch sizes, or rates, lead times on deliveries, etc. The implications of such uncertainty are evident in two main respects. Firstly, uncertainty about the supply of any resource will probably encourage management to provide adequate protective resource stocks, i.e. to insulate the function from uncertainties or fluctuations in the supply environments by exercising some choice of structure (structure choice and internal factors will be examined below). If, however, for other reasons input stocks are infeasible, i.e. because of uncertainty of the nature of demand, this strategy is not available. In such cases it is likely that the resource will be ordered or requested to satisfy specific customer demands, following which the system must await the arrival of such resources. In other words little or nothing can be done until the input is available. We can consider such a system as a multi-input channel customer pull system in which there is no provision for stock in at least one input channel. The system is in effect limited by *an* input. There is some similarity with input push systems, i.e. structure SCO or DQO. However, there is likely to be some, albeit incomplete, control over input flow.

Case examples

Motor vehicle sales

The UK company of a major international motor manufacturer introduced a new model in 1969. It represented a new departure in marketing, in that a considerable range of builds, specifications and options was offered. Five different engine builds were available. Additionally, five basic bodywork specifications were made (two of which were available only with one type of engine specification). There were a total of 26 basic engine/body derivatives for the customer to choose from. Taking account of features such as trim, tyres, gearbox, but ignoring colour specification, radio, etc, there were over 800 models to choose from. The company claimed that they could build $1\frac{1}{4}$ million such cars without any two being precisely the same. This situation had major implications for the retailer. If the customer was to be encouraged to exercise the choice available to him, the retailer would find it impossible to hold stocks. He would therefore be obliged to supply goods on a 'from source, direct to customer' basis, i.e. structure DOD. Such an arrangement would be unusual for a vehicle retailer who normally operates with a showroom full of new vehicles to tempt his customer, i.e. structure SOD. Increase in the range of items to choose from inevitably reduces the ability of the retailer to predict the customer's demands, and therefore eliminates the possibility of an operating structure which requires input stocks (of goods). Similar implications apply for the manufacturer. The adoption of such a model strategy greatly increases the uncertainty of the nature of customer demand. This in turn makes the holding of finished goods stocks impossible, and hence it becomes impossible to manufacture to stock in anticipation of demand (i.e. structure SOS). However, even in such circumstances it remains possible to hold stocks of most input materials since they will be used in the manufacture of a variety of models. Hence despite uncertainty of customer demand, there remains some predictability of input requirements. Hence structure SOD might be likely to apply.

Naturally this situation is impossible. Neither suppliers nor manufacturers of low cost family cars would *choose* to operate in this manner, nor did they for in effect customer choice was limited to availability. At a later date the company substantially reduced the number of specifications and options available to the customer.

Management education

A small college had for many years successfully run a two-month course for senior business managers. The course was residential, ran several times yearly, and attracted large numbers of applications from potential participants. The college employed teaching and administrative staff for this programme, had all necessary resources and had no dependence upon outside resources. It had been the principal activity of the college for some time, but in recent years demand had begun to fall off, and the college began to introduce other courses. Eventually in order to sustain a sufficient level of course activity to meet costs, and also in order to take advantage of changing ideas and needs in management education, the college moved almost entirely into the business of designing and running courses for organizations on a 'one-off' basis, supplemented where necessary by conferences, the letting of facilities, etc.

Such developments completely altered the nature of the operation of the college. Whereas previously the nature of customer demand was entirely predictable, this was no longer the case. Previously the system had operated 'from stock, from customer stock' (i.e. structure SQO) that is with a waiting list of customers for its one service. It had then become a 'from stock, direct from customer' operation (structure SCO). It was now moving towards a 'from source, from customer stock' operation (structure DQO), that is, the nature of customer demand was identified and resources were then acquired to satisfy this demand. In other words because of the variety of possible courses and activities offered it was no longer possible to stock all necessary input resources. In respect of teaching staff resources in particular, structure DQO applied, and in some cases other resources, e.g. equipment and facilities, had to be acquired for certain courses.

Glass manufacture

An entirely new process for the manufacture of flat sheet glass had been developed by the major UK glass manufacturing company. The process enabled glass to be produced in a variety of widths and thicknesses at a far greater speed than previously possible. The company dominated the UK market, a situation which facilitated the introduction of a new marketing policy shortly after the widespread adoption of the new process by the company. Previously the company had supplied a range of over 100 sheet sizes to customers. It was not practical to hold stocks of this range of sizes, hence in most cases glass was held in stock in large sheets and subsequently cut to one or more of the smaller sizes to satisfy specific customer orders.

This range was reduced to approximately 25 standard sizes. Most customers were required to order and take glass in these sizes. These customers then often cut smaller sizes as appropriate or as required by their own customers. This rationalization of the product range enabled the glass manufacturer to introduce substantial *final* goods stocks which in turn facilitated the provision of a good customer service in respect of deliveries. (This 'ex-stock' service was seen as one benefit to the customer to offset the reduced size range.) The feasibility of manufacturing 'from stock, to stock, to customer' (structure SOS) in fact stemmed from the increased predictability of customer demand, in turn created by the restriction of customer choice. (A situation similar to the eventual outcome of case 1 above.)

INTERNALLY RELATED FACTORS

External factors will rarely, by themselves, determine structure. Other factors will have some effect on system structure. We can consider such factors to have some *internal* relevance in that they are within the direct control of the operations manager. We can illustrate the influence of such factors by examining how the structure of systems might change or be changed. Structure change might take place over time when no appropriate internal action is taken by management to prevent it. Alternatively management might take internal action in order to bring about change; in other words the manipulation of some internal factors provides a means for management to exercise some *choice* of system structure.

The following examples illustrate two aspects, which for convenience we call *change* (i.e. occurring when no internal action is taken) and *choice* (i.e. brought about as a result of the manipulation of internal factors). The two are obviously related both in respect of cause and effect.

Service

Change

The resources utilized in a dentist's surgery (e.g. dentists' chairs, equipment, etc) will have been acquired in quantities which are sufficient to more or less match or accommodate the demand expected to be placed on them. Alternatively, given the availability of certain resources, the service will have undertaken to deal with a given number of customers, again the intention being to match demand with available resources. Given this balance the operation of the system will usually involve a customer appointments procedure, customers being asked to arrive at the rate approximately matching the service rate, typically around four or five customers per hour. It is intended therefore that the system should work in the manner of structure SQO, customer waiting being tolerated, and provided as a means to ensure good resource productivity. However such a service system is rarely entirely predictable. Customers might arrive late or early and service times will be variable. Furthermore demand may rise or fall over a period of time. If the latter the system will in effect change to structure SCO. If the former, structure SQO will remain but of course customer service will deteriorate.

Certain service systems will ideally operate with structure SCO. A hospital accident ward should ideally operate in this fashion; however in the event of demand exceeding the capabilities of the available resources the structure will inevitably change from SCO to SQO.

Choice

In some cases such changes might be introduced as a matter of policy in order to provide for improved resource productivity. The choice between for example structure SCO and structure SQO will be influenced by the relative priorities of customer service and resource productivity. In certain cases, e.g. the hospital accident ward, the policy will be to operate the system as structure SCO without customer waiting. However, in other similar situations, e.g. general medical or general surgery wards of hospitals, whilst ideally structure SCO might be utilized an operating policy involving structure SQO will often be adopted.

Changing circumstances, for example the cost of resources and facilities or the nature of demand, may give rise to the choice or decision to change the systems in this manner. Equally in establishing systems such a choice will exist.

Manufacture

Change

Given a knowledge of the nature of customers demands, i.e. given the existence of a market for a certain product and therefore the feasibility of providing a finished stock of goods, a manufacturing system may be designed such as to provide an output rate approximately equal to the average demand rate for its goods. Resources would be acquired, e.g. machine tools, labour, etc, to provide for this required average output rate. In the short to medium term therefore the output rate will be fixed. However, demand levels may well fluctuate as a result of, say, seasonal factors or other disturbances. The existence of the stock of goods will help decouple customer and function, thus enabling a relatively steady output rate to be maintained and thus relatively good resource productivity achieved despite the fluctuating demand.

In such circumstances, should either average demand rate increase or average output rate fall, the output of stocks will eventually be starved and structure SOS or DOS will change to structure SOD or DOD respectively. An increase in demand rate might result from product price changes, increase in market size, the actions of competitors, etc, whilst changes in output rate might result from changes in the level of resources utilized or available in the function, e.g. deteriorating equipment, labour turnover, absenteeism, stoppages, etc. Thus for reasons perhaps beyond the control of management, structure changes may occur.

A similar situation may occur in respect of input resource stocks. Many manufacturing systems will operate with structure SOS. The replenishment of input stocks will be arranged such as to occur at a rate approximately equal to the rate of usage of such resources by the function. An increase in the usage rate or a fall in the supply rate will lead to the exhaustion of input stocks and therefore a change from structure SOS to structure DOS. An increase in usage rate may result from a management decision, however a reduction in supply rate caused by, for example, disruptions in suppliers, rejection of faulty resources, etc, might be beyond the control of management, and thus change in structure from SOS to DOS might occur for external reasons.

Choice

In some cases, particularly where alternative sources of goods are not readily available to the customer and where customer service is not for this or for other reasons of paramount importance, structure SOD may be introduced as a matter of policy in preference to structure SOS. In such circumstances customers will be required to wait for the provision of goods. A decision to restrict output, i.e. to provide an output rate less than the average demand rate will have precisely the same effect. Such changes might be introduced or alternatively such a choice might exist at the time of establishing the system.

Transport

Change

An emergency or accident ambulance service will often be designed and intended to operate without customer queueing, i.e. in the manner of structure SCO. The resources required by the system (e.g. ambulance, drivers, service facilities, etc) will be acquired such as to at least match and possibly exceed expected demand rate for the service. An increase in demand rate, however (caused perhaps by a major accident, population growth, etc, or a reduction in the amount of resources available caused perhaps by vehicle breakdown, staff sickness, disputes, stoppages, etc), will give rise to the existence of structure SQO in which customers are required to wait. Such changes might therefore occur for external reasons beyond the immediate control of operations management.

Choice

In certain cases, for example the design of an ambulance service for patients requiring non-urgent hospital attention, e.g. out-patients, structure SQO may be preferred to structure SCO. In such cases, furthermore, structure SQO may be practical and since customer queueing might be tolerated and indeed expected, a reduction in customer service may be traded off against an increase in resource productivity. In such circumstances the opportunity to change systems might occur, equally the choice will exist in creating systems.

Choice and change

These examples indicate the manner in which structures might change, be changed or be chosen. It is clear that in many cases a choice of structure will exist. Such *choice will of course be limited by questions of feasibility*, since when the necessary prerequisites for a structure do not exist management will have no opportunity to adopt that structure. Thus for example in manufacture, given predictability of the nature of demand, management might select from SOS, DOS, SOD and DOD according to their policy and requirements. Without this prerequisite for structures SOS and DOS the choice will be between SOD and DOD as the only two feasible structures. Notice that in selecting a structure one of the principal considerations will be the relative importance of customer services and resource productivity.

Given the feasibility of certain structures the choice between them may reflect management's view of what the customer wants or will be prepared to accept by way of service, and the need within this constraint to maximize resource productivity. The degree of *urgency* associated with the customer's demand is often of importance in the choice of structure. It is one of the main reasons for the existence of the structures SOS and DOS. A cafeteria is essentially structure SOS, whilst a part of a restaurant (that part which prepares food) is normally structure

SOD. The former provides the fastest service, whilst the latter requires the customer to be more patient. Whether structure SQO or SCO is chosen will largely depend upon the degree of urgency associated with customers, i.e. the time they are prepared to wait. In customer push systems, it is generally true that the degree of urgency is less when the customer's items or goods are input, rather than his person. Further, the length of time the customer is prepared to wait is generally related to the time required to complete the function, e.g. the public may be happy with the service provided by four or five trains operating on an inter-city route with a journey time of several hours, yet they would be quite unhappy with such a frequency for a short commuter service.

Structure changes of the type illustrated by the above examples, arise largely from changes in the balance between demand and the capability of the systems. Similarly structures are often changed, i.e. a choice is exercised by the manipulation of system capability relative to demand. A relative increase in the level of customer demand can give rise to the depletion of stocks. A relative decrease in the level of customer demand can give rise to the creation of stocks. Loss of balance will often therefore give rise to changes in structure. The same changes will result whether or not changes in balance occur as a result of demand level changes or changes of function capability. These changes are therefore related to the question of capacity—*function capacity*. If capacity is changed or changes relative to average demand levels, a structure change may result. We can therefore consider this source of influence on system structures to be internal, since it is within the scope of the control of operations management. Management can change structures by changing capacity, can allow changes to occur by not adjusting capacity to balance demand level changes, or can avoid structure changes by manipulating capacity to maintain a balance with a changing demand level. We shall shortly examine the significance of such effects, particularly structure *changes* of the type illustrated in the above examples, i.e. those which management allow to occur. Suffice to say that at this stage such changes may well affect resource productivity and customer service, but *may* not affect the manner in which the system is managed.

Exhibit 5.2 summarizes the nature of the internal and external factors influencing system structure. The nature of the function and the influence of the customer (i.e. whether customer push or pull) will determine which of the seven basic structures are *appropriate*, whilst the predictability of the nature of demand will influence *feasibility*. Management choice or *preference*, based mainly on consideration of customer service and resource productivity objectives will give rise to the adoption of one feasible structure. This structure might subsequently *change* or be changed mainly as a result of a change in demand relative to capacity.

Change and time

The question of structure change leads us to consider the influence of *time*. We have seen that structure change can take place over a period of time, and have

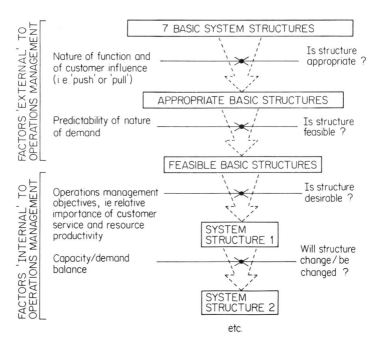

Exhibit 5.2 Factors influencing system structures

noted (Chapter 4) that for this reason we should consider the structure of systems at a given point in time. Most systems will normally have a particular structure. For example, a retail shop will normally be structure SOD, whilst a motel will normally be structure SCO. Shortly we shall consider the problems associated with the management of given structures. The implication is that the nature of management problems will differ between structures. If a system is designed to have, or if it must normally have, a certain structure the strategies adopted by management should reflect the needs and constraints of that system structure. If the structure changes for whatever reason, either the approach to the management of the system will remain basically unchanged or the approach will be changed. It is likely that in the case of temporary structure changes the approach to management will not change, but in the case of long-term change a corresponding management change will be desirable. Clearly a system may not always exist or work as it is intended to. For example all customers must ultimately be prepared to wait or queue in a push system. Even an ambulance service may require customers to wait. However, it may be inappropriate or unnecessary to consider the system structure to change every time such a queue forms. The length of time customers are prepared to wait is perhaps too short to permit the system to be run or managed in a different way. The way the system is managed may not therefore change at all, despite a temporary structure change. In practice the manner in which a dentist manages will not change simply because the non-arrival of a few patients causes the system temporarily to change from SQO to SCO. Short-term structure changes may affect customer service (e.g. the ambulance service changing from SCO to SQO) or resource productivity (e.g. the dentist changing from SQO to SCO) but little else.

At the risk of complicating matters slightly it is worth pointing out at this stage that *perceptions* of system structure may differ. The operation of major car ferries such as those across the English Channel or North Sea is an interesting case in point. It may well be that the operators of such ferries see them operating in the manner of structure SCO. That is, customers are booked onto a particular ferry and are required to arrive at the appropriate time for that ferry. Customers, however, may see the system as having structure SQO, since they may consider themselves to be waiting for the ferry. In fact, if such waiting takes place, it is meaningful to the customer but not normally to the ferry operator, simply because their operational time spans are different. The fact that cars are waiting on the quayside is significant for the customer since he may feel that a wait of half an hour or so is a significant part of his one-week holiday. The fact that such cars are waiting is not, however, significant to the ferry operator since it does not afford him the opportunity to operate the service in any different manner. He therefore perceives the system to have structure SCO for all practical purposes, and therefore manages the system accordingly. It is his perception which is of relevance to our present discussion. We are concerned mainly with system structure as it is seen to exist or assumed to exist by operations managers.

The possibility of structure change will not therefore inevitably affect our discussion of management problems, providing we consider the normal or actual system structure, as appropriate, and providing we bear in mind that if systems are susceptible to frequent change a compromise approach must be adopted by management, i.e. one which is appropriate for all likely structures yet not necessarily the best for any one.

In the remainder of this book we shall examine the problems encountered and procedures adopted by operations managers. We shall look at the management of each type of system, concentrating as pointed out in Chapter 2 on those aspects which distinguish systems. For the reasons given above we shall not presume that, for example, a change in system structure necessarily gives rise to the adoption of a different approach by management. However, we shall base our discussion on the examination of particular structures and will continue to look at structure change.

SUMMARY OF CHAPTER 5

There are certain external *factors which influence the feasibility of certain structures*. The principal factor is the extent of *predictability of the nature of demand* on the system.

Within the feasibility limits determined by external factors system structures can *change* and/or be *chosen* by operations managers.

Such structure changes/choice derive largely from the variation of *internal factors* in particular *system capacity*.

Structures may change over *time*; and individuals' *perceptions* of structure may differ. We need only be concerned with the structure which is seen to exist, or assumed by operations managers, i.e. the structure which influences the operations managers' approach to the management of the system.

Section 2

Operation Management Problems

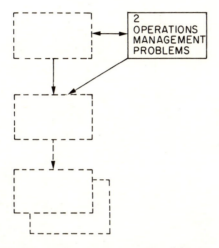

CHAPTER 6 Operations Management Problems

In this section we will identify those operations management problem areas which distinguish systems, i.e. those problem areas whose characteristics are influenced by system type.

CHAPTER 6

Operations Management Problems

The decisions of operations managers can affect system structure, that is within given feasibility constraints management can select and/or change structures. This chapter will explore the relationship of structure and problem area. We have argued that to some extent the nature of the problems facing operations management is a function of system structure. It is worth repeating that it is not which structure exists at a point in time that really matters, but which structure normally exists, or is assumed by management to exist, or is intended to exist. Whether a system temporarily runs out of stocks or accumulates a queue of customers is important in respect of customer service and/or resource productivity, but it is of direct relevance to the nature of the management of the system *only* if management change their mode of operation as a result.

In this chapter we shall establish a foundation for an examination of the nature, and practice of operations management. In subsequent chapters we shall examine particular aspects of operations management and consider the influence of system structure. We shall examine each of the important operations management problem areas in turn. Our task in this chapter therefore is to identify these problem areas.

We shall examine only the principal problem areas of operations management, i.e. those in which decisions made and actions taken can influence system structure and where the type of decision and action required can be influenced by structure. Such problem areas therefore are *characteristic* of a system structure. Their nature and complexity will be influenced by structure. Thus in some operating systems such problems may be relatively easily solved whilst in others their solution may be more difficult. In some situations certain methods, procedures or techniques might be appropriate in tackling a problem, whilst in others they might be quite inappropriate. Insomuch as the nature and character of operations management might differ, such differences will be reflected through the nature and characteristics of principal problem areas. Insomuch as operations managers are able to influence the nature of the operating system and thus the nature of the organization they will do so through their decisions in these principal problem areas.

It follows that the nature and characteristics of the problems in the second set are in most respects *common* to the seven basic structures. Irrespective of the type of system structure we would expect to find similar types of problems, which might be tackled in much the same way, using the same procedures, etc. As far as these problem areas are concerned, the nature, character and complexity of operations management will be similar irrespective of the type of organization, function or industry. This does not imply that such problem areas are less important to the successful operation of the system. Correct decisions must be made irrespective of the area. However, these common problems *may* be considered to be more readily dealt with, if only because of the transferability and relevance of methods and procedures. In Chapter 2 we noted that one argument advanced, to demonstrate that the management of operations might be considered as a single subject area, was the existence of apparently similar problems in ostensibly different types of system. It was noted that certain decision making techniques and procedures were therefore applicable in different types of system. We shall show that to a large extent these 'similar' problem areas are what we have now referred to as *common* problems. The existence of these alone is an insufficient basis for accepting operations management as a single subject drawing upon a common body of knowledge. The development of the subject requires also an adequate understanding of the nature of the characteristic problems. Furthermore the existence of operations management as a subject or management function at a *policy* level requires an understanding of these principal problem areas, through which management can influence the structure and hence the nature of the operation and the organization.

PRINCIPAL PROBLEM AREAS IN OPERATIONS MANAGEMENT

Inventories

The management of the physical stocks or inventories in operating systems is clearly a function of system structure, if only because certain structures provide for the existence of stocks whilst others do not. The location of inventories is a function of structure, as also is the nature of the inventory management problem. Defining inventory management as the planning and control of physical stocks, both aspects of the problem may be affected by structure and may also affect structure.

Some predictability of the *nature* of demand will permit the provision of inventories. Such inventories will then often be provided in order to protect or insulate functions from demand *level* fluctuations. The provision of inadequate stocks may give rise to a change in structure through loss of inventory. Thus, whilst in capacity management an objective will be to match capacity with the long-term demand level to avoid a structure change, the inventory management problem is complementary since an objective there will be to accommodate short-term imbalance again to avoid a structure change. In respect of input stocks an objective will be to protect the function from fluctuations in levels of supply. The manner and ease with which this might be achieved will be influenced by the nature, i.e. the structure, of the remainder of the system. The nature of the inventory management problem is therefore *characteristic* of system structure and may be considered a principal problem area of operations management. The methods, techniques and procedures deployed in tackling inventory management problems will be influenced by system structure, and the success achieved in tackling this problem may in turn influence system structure.

The physical nature of an operating system will largely reflect the nature and location of its inventories, and the management of such inventories will influence both resource productivity and customer service. The existence of output stocks may facilitate the provision of high customer service at least in terms of availability or 'timing' (see Exhibit 3.4). However, their existence may be costly, a fact which may be reflected in the cost and choice aspects of customer service. The provision of input resource stocks may in certain respects benefit customer service, yet resource productivity may be adversely affected because more resources are idle.

Exhibit 6.1 illustrates the nature of the inventory management. For output stocks, information will be received generally from customers in the form of orders. This will in turn give rise to requests for stock replenishment, which will be 'input' to activity scheduling and control. Information relating either to customer requests or output stock replenishment will be received into input

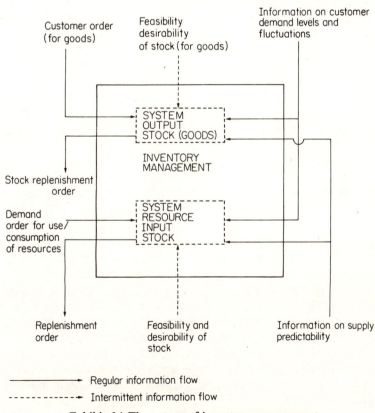

Exhibit 6.1 The nature of inventory management

resources stock, where these exist, through activity scheduling and control and these in turn will give rise to requests for input stock replenishment placed on suppliers. Information on customer demand levels and fluctuations in demand, will be required intermittently in the determination of stock policies.

Scheduling problems

There was some indication in the discussion in Chapter 5 that the nature of some aspects of *scheduling* in operations management are characteristic of system structure and that decisions taken in scheduling can affect structure. Scheduling is concerned with the timing of occurrences. *Operations scheduling* in its widest sense may therefore be considered to be concerned with the specification (in advance) or the timing, of occurrences within the system, of arrivals to and departures from the system including arrivals to and departures from inventories within the system. Thus we can consider the problem of inventory management as described above to be a part of a wider overall scheduling problem in operations management. The nature and extent of this overall scheduling problem will therefore be influenced by the presence and location of inventories and the relationship of the customer to the system, all of which are characteristic of system structure. As with all characteristic problems, the procedures and methods deployed in scheduling may be influenced by structure, and the effectiveness of scheduling may in turn affect structure. For example the opportunity to schedule customer arrivals in input push systems and the effectiveness of such scheduling might enable a system to operate without customer queueing, i.e. structure SCO rather than SQO. Conversely ineffective scheduling of arrivals in structure SCO systems might give rise to a change to structure SQO with consequent deterioration of customer service and ultimately loss of trade. In some demand pull systems it may be possible to schedule activities within the system to provide for outputs at times which will coincide with expected occurrence of demand. In such situations the need to store output may be avoided (i.e. structure SOD or DOD rather than SOS or DOS). Conversely failure satisfactorily to schedule outputs in such systems may give rise to a structure change.

Whilst in structure SOS where both input and output stocks exist, operations scheduling might necessitate the examination of the three stages of the system, i.e. the timing of inputs to stock, to function, and to stock, other structures would seem to provide a simpler problem since fewer stages are involved. On the other hand, in such structures the relationship of the customer and the function, and the supplier and the function, may be direct, hence somewhat different and perhaps more complex scheduling problems might exist.

If we consider operations scheduling to relate to the physical flow or transfer of resources or goods then the nature or extent of the overall scheduling problem is clearly influenced by the number of stages involved in the system, and therefore by structure.

In many situations a clear distinction is made between operations planning and control. In manufacture, for example, production planning (or scheduling) and production control are often seen as two separate functions. In fact, however, as we have seen in Chapter 4 (Exhibit 4.5) planning (or scheduling) and control are mutually dependent. We shall therefore consider operations scheduling and control as one aspect or problem area in operations management. We have shown that scheduling problems can influence and be influenced by system structure. It follows therefore that the nature of the problems of *control* is a function of structure. The number of control 'loops', in system terms, will reflect the manner in which schedules are applied. Effective control is necessary for effective scheduling and vice versa. Scheduling decisions may be ineffective

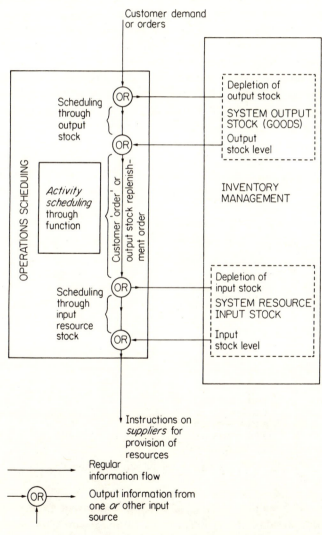

Exhibit 6.2 The nature of operating scheduling

when inadequate controls are applied to ensure that activities and events occur as scheduled. Thus the structure changes described above as resulting from scheduling decisions may not in fact result without the use of effective control. The effects of scheduling failures identified above might also be associated with control failures. Structure changes may result from the adoption of certain scheduling and control decisions, and failures in one or the other might also give rise to structure change.

Exhibit 6.2 illustrates the nature of operations scheduling and shows the role of inventory management in the scheduling of occurrences in operating systems.

Where output stocks exist, customer demand will be met by scheduled output from stock, such stocks being replenished by scheduled inputs. In the absence of such stocks, customer demand will be met by scheduling output from the function which in turn will necessitate the scheduling of resource inputs either from input stock or direct from suppliers; if the former then input from suppliers must be scheduled for stock replenishment.

Exhibit 6.3 takes a somewhat narrower and perhaps more conventional view of the scheduling problem. *Activity scheduling* is an aspect of operations scheduling and is concerned only with flow through and the timing of occurrences in

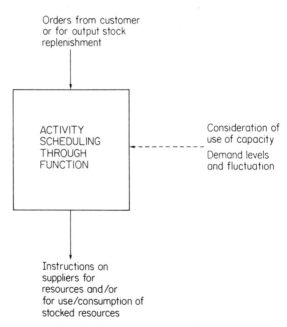

Exhibit 6.3 The nature of activity scheduling

relation to the function. Activity scheduling and inventory management together provide scheduling decisions for the entire system. They are in effect the two complementary aspects of operations scheduling.

Capacity

Changes in system capacity relative to demand have been shown to relate to, or be reflected in, changes in system structure and/or to affect efficiency in customer service and resource productivity when operating as for a given structure. The determination and adjustment of capacity in an operating system is therefore an important problem area, since decisions taken in this respect may intentionally or inadvertently change the nature of the system and/or affect the efficiency of operation of a particular system. Equally, decisions not taken or wrong decisions may also result in structure changes and/or loss of efficiency as for example following the failure to adjust system capacity to match customer demand changes. The planning and control of capacity is both important and complex, and furthermore the nature of the problem will often be affected by structure. In other words for a given system structure the capacity management problem may well differ from that facing management operating in a different structure. Since structure affects the nature and complexity of the capacity planning problem, the methods, procedures and techniques appropriate for tackling the problem may also be influenced by structure. In all respects therefore we can consider capacity management to be a principal problem area, the nature of which is *characteristic* of system structure.

The management of system capacity is of crucial importance in operations management. The determination of capacity is the key system *planning* or design problem and the adjustment of capacity is the key problem area in system *control*. Capacity decisions will have a direct influence on system performance in respect of both criteria, i.e. resource productivity and customer service. It is difficult to see how any organization can operate satisfactorily without good capacity management. Excess capacity inevitably gives rise to low resource productivity, whilst inadequate capacity may mean poor customer service. Decisions taken in other areas are unlikely to offset errors in this area.

The capacity problem is often of a medium to long-term nature. Since system capacity is a reflection of the nature and amount of resources available in the system, short-term adjustments are often impossible. Capacity management is primarily concerned with the matching of resources to demand. It is concerned therefore with the levels of resources and demand. It is worth noting that whilst the *nature* of demand, in terms of what the customer requires, may influence structure, or rather determine the feasibility of structures, the success achieved (through capacity management) in accommodating demand in terms of *how many* the customer requires may also influence structure. It is only in this respect that demand level will influence the structure of operating systems. High or low demand in itself will not influence system structure but demand which is high or low relative to available capacity, will. Furthermore, since the difficulty of matching capacity to demand will in part depend upon the timing and the extent of the changes required, the nature of the changes in demand levels may also influence system structure.

Exhibit 6.4 illustrates the nature of the capacity management. Decisions in respect of system capacity will affect inventory management both through their

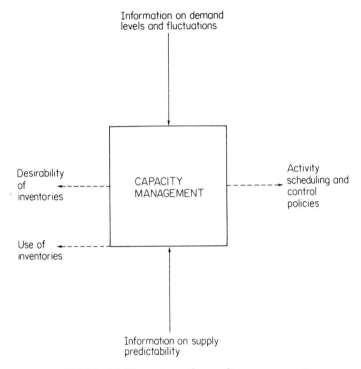

Exhibit 6.4 The nature of capacity management

influence on the choice or the desirability of the use of inventories and also the manner of the use of such inventories. Capacity management decisions will also influence activity scheduling. Capacity management decisions will largely reflect assessments of customer average demand levels and fluctuations in demand, and will be aimed at matching system resources to such customer needs.

OTHER PROBLEM AREAS

Inventory management, activity scheduling and capacity management are, using our terminology, the principal problem areas of operations management. It follows that those other problem areas identified in Chapter 2 can be considered to constitute a different class. Such problem areas as facility location and layout, quality control, maintenance and reliability, job and work design, can be considered to be *common* problems, that is, they have no significant and direct relationship with system structure. Decisions taken in each of these areas are unlikely to have a direct effect on system structure, and conversely structure will not directly influence the nature of such problems. It is likely therefore that in each such area, any methods and techniques which are available to facilitate management decision making will be of some relevance irrespective of structure. In other words, these are the common denominator problem areas where skills and techniques are more readily transferrable between situations. Such areas, problems and techniques tend to be cited as evidence of the similarity of the management of different types of operating system. Problems of location, replacement, etc, occur in manufacture, transport service and supply, and similar decision-making procedures are relevant in each type of system. In contrast the nature of the problems in each of the three principal areas is a source of distinction between operations management in different systems.

In our examination of the nature of operations management we shall deal exclusively with problems of scheduling and control, inventories and capacity. These are complex problems and central to effective operations management. *They are of direct relevance at a policy level in an organization.*

INTERRELATIONSHIP OF PRINCIPAL PROBLEMS

One factor adding considerably to the complexity of inventory, capacity and activity scheduling problems is their close interdependence. Decisions taken in one will have a direct impact on performance in the others. Such interdependence is less evident in the other problem areas, a fact which tends to 'underline' the central importance of these three problem areas in the management of operations. The nature of this interrelationship was identified in the preceding discussion and in Exhibits 6.1 to 6.4. In many respects the problems of inventory management and scheduling are subsidiary to the problem of capacity management. Capacity management decisions will determine how the operating system accommodates customer demand level fluctuations. Capacity management decisions will provide a context within which inventories and activities will be both planned and controlled. Capacity management decisions will to some extent reflect operating policy decisions whilst inventory and scheduling problems might be considered as more tactical level issues. In other words capacity management is a medium to long-term problem area, whilst inventory and scheduling problems are short to medium-term.

Exhibit 6.5 combines Exhibits 6.1, 6.2, 6.3 and 6.4 and provides further details to indicate the nature of the interrelationship of the three principal problem areas. There is a relationship between the problems of inventory management and those of activity scheduling and control. Where both input and output stocks exist, activity scheduling will provide a link between the two.

Where only output stocks exist activity scheduling will provide a link with resource suppliers, to provide stock replenishment. Where only input stocks exist the link with customers will be provided through activity schedules. Within the constraints of feasibility, capacity management considerations will influence the deployment and use of inventories. Not only will the decision to provide stocks reflect capacity management policy for the accommodation of customer demand level fluctuations, but also the management of such stocks will be influenced by capacity considerations. For example, stock reorder policies and levels will be determined in the light of capacity considerations.

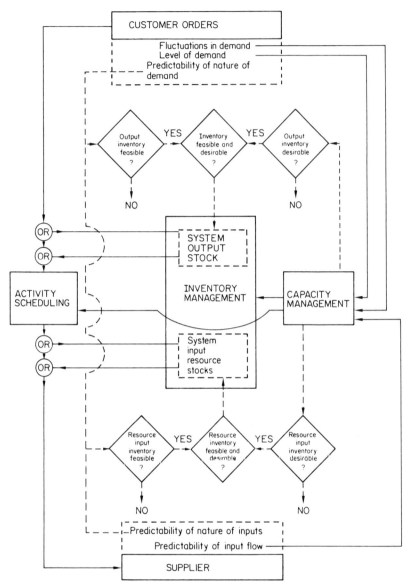

Exhibit 6.5 The interrelationship of capacity management, inventory management and activity scheduling

Exhibit 6.6 indicates the nature of the interdependence of these three principal
problem areas in operations management. Each of these three problem areas will
be examined in more detail in the following three chapters.

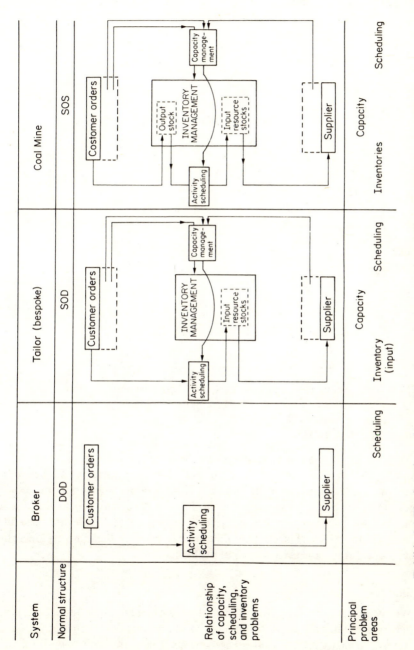

Exhibit 6.6 The interdependence of the three principal problem areas of operations management

EXAMPLES

The capacity, scheduling and inventory characteristics of three of the systems listed in Chapter 3 are examined below and are summarized in Exhibit 6.7.

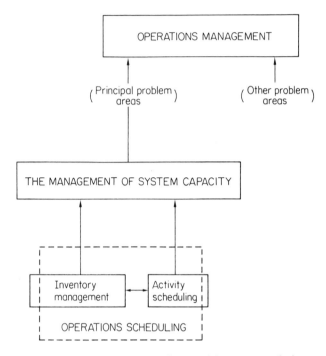

Exhibit 6.7 Capacity, scheduling and inventory relationships—examples

Broker (structure DOD)

Here, as we have seen, a relatively simple customer pull structure exists, i.e. 'function from source, direct to customer'. No inventories exist, hence the inventory management problem is absent. Furthermore, since no input resource stocks exist the capacity management problem is effectively absent. In practice therefore orders are received directly from customers and these are scheduled directly on to the supplier(s). This therefore is perhaps the simplest operating system as regards these three principal problem areas.

Tailor

We shall consider a bespoke tailor, making garments to individual customers' orders using materials and resources on hand and in stock. Structure SOD will therefore normally apply. In this case output stocks are both undesirable and infeasible. Input stocks are feasible since there is normally sufficient predictability of the nature of demand to permit materials, machinery, etc, to be provided in advance. Input stocks are also normally desirable since, given the lead times required by customers, fluctuating demand levels cannot be satisfied by placing orders directly on suppliers. Furthermore, resource input inventories may also be desirable in view of some unreliability in supply, in particular material supplies. The task of capacity management will be to monitor demand levels and fluctuations and to apply a strategy for capacity planning and control to influence both inventory management and activities scheduling decisions. Orders will be received directly from customers and these will be scheduled to the function, in other words activity scheduling will communicate customers' orders in a suitable form through to input resources stocks. Replenishment of these resources will in turn lead to placing demands on suppliers. The principal problem areas in this case therefore are scheduling, inventory (input) and capacity. Inventory management decisions in respect of system input resources, e.g. level of resources carried, clearly affects the capacity of the system. Equally, decisions in activity scheduling will also have a bearing on system capacity and the capacity management problems.

Coal Mining

Here the normal structure provides for both input and output stocks and hence system structure SOS will apply. In this case output stocks are feasible since the nature of demand is fairly predictable. System output stocks are also desirable since the demand levels will be subject to some degree of fluctuation because of seasonal factors in the demands of coal stocks. The resource input inventories are both feasible and desirable. The feasibility derives from the predictability of the nature of demand, hence all the necessary resources, e.g. labour, materials, machines, etc, can be provided. Resource input inventories are desirable in view of the need for continuous steady production, coupled perhaps with the likelihood of some uncertainty of supply.

Orders will be received from customers into output inventories. This, through inventory management, will lead to the placing of stock replenishment orders, or the requesting of a continual flow to stock at a given rate. This replenishment need will be met through activity scheduling decisions acting upon the resources in the system, which will in turn be replenished from supplies. Capacity management will monitor demand levels and fluctuations and supply reliability, seeking to establish and maintain a capacity strategy appropriate to accommodating the uncertainty surrounding the system. Capacity decisions will affect policies in inventory and scheduling.

SUMMARY OF CHAPTER 6

Certain *principal operations management problem areas* can be identified, i.e. those in which decisions can influence system structure, and/or where system structure influences the nature of the problem. These can be considered as problem areas which are *characteristic* of system structures.

Other problem areas can be considered as *common*, i.e. their nature is largely unrelated to system structure.

Principal or *characteristic problems* concern system *capacity*, activities *scheduling* and *inventory* management. Activity scheduling and inventory management together comprise the operations scheduling function for systems.

Capacity, scheduling and inventory problems and decisions are of direct relevance at a *policy* level in an organization.

Capacity, scheduling and inventory problems are closely *interrelated.*

Section 3

The Problem Characteristics of Operating Systems

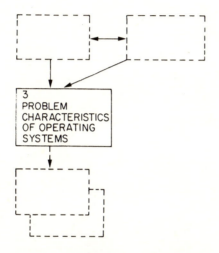

In this section we will consider the problem characteristics of systems through an examination of the nature of the three principal problem areas of operations management. The type of problem which can occur, the influence of system structure and the operations management strategies which might be employed will be considered.

CHAPTER 7

The Nature of Capacity Management

The effective management of capacity is perhaps the most important responsibility of operations management. Not only may decisions taken affect the nature of the operating systems, they will also have a significant impact upon the nature of the two other principal problems, i.e. activity scheduling and inventory management. The management of system capacity is both complex and difficult. Numerous techniques and methods are available; however the sheer complexity of capacity problems prevents these being of value other than in tackling small aspects or perhaps rather abstract versions of the problem. It is appropriate that before examining such decision making techniques, we look at the problem of capacity management both in general and for each of the seven basic structures. In this chapter therefore we shall look at the *nature* of capacity management, the various approaches to the management of capacity, and their relevance to the seven basic system structures. In Chapter 10 we will outline the procedures and methods available for decision making in this area and relate these to the approaches which might be appropriate in each structure.

THE CAPACITY MANAGEMENT PROBLEM

We have indicated that the objective of capacity management, i.e. the planning and control of system capacity, is to match or balance the level of operations to the level of demand. We have further indicated that it is the uncertainty of demand level which gives rise to this problem. The existence of a stable and known demand would considerably simplify the problems of capacity management. A changing demand situation can be accommodated relatively easily providing such changes can be accurately predicted. The expectation of changing demand levels without the possibility of accurate prediction gives rise to extremely complex capacity problems. We should perhaps note that this uncertainty of demand level may be caused by:

(1) Uncertainty of the number of orders to be received, and/or
(2) Uncertainty about the amount of resources required for the satisfaction of customer orders.

We have pointed out in Chapter 4 that in some cases the demand placed upon a system can be influenced by those within the system. In such cases demand levels might not be seen simply as *given* constraints. It is an oversimplification to suggest that management seeks only to manipulate resources within a system to satisfy given external requirements. However, we have argued that the operations manager alone will normally have insufficient direct influence to enable him to control demand to such an extent as to appreciably affect his need to plan and control system capacity. Nevertheless, whilst he might not control demand, he may choose to ignore some of it. He may, for example, choose to attempt to satisfy only a few of his potential customers. At the other extreme, an approach might be adopted in which the objective is to satisfy all potential demand. Most systems will seek to operate between these two extremes, the approach adopted being influenced by capacity management considerations as well as being a factor influencing capacity management. Thus, in situations where changes in capacity are not readily accomplished, there may be some benefit in attempting to satisfy only a small portion of total potential demand, and vice versa.

In most situations management of system capacity will be influenced by the level and fluctuations of demand, or more precisely that portion of total potential demand which is recognized as applying to the operating system at that time.

Capacity management is a medium to long-term problem area. Exhibit 7.1 locates the two elements along such a timescale. Taking our two elements of operations management, i.e. design and planning, and operation and control, the former is seen to be largely a 'future' activity, and the latter essentially a 'present' activity in this, as indeed in other, aspects of management. Capacity planning is considered as an aspect of medium term planning whilst operation

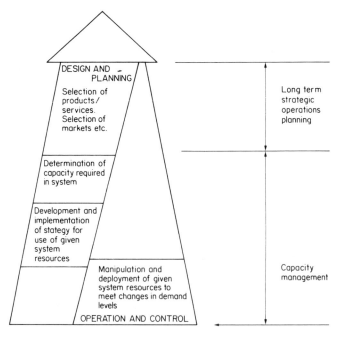

Exhibit 7.1 The timescale of capacity management

and control consisting of the implementation of capacity plans is short to medium term.

Capacity planning will involve the study of likely demand patterns for the medium to long term, the determination of the capacity required to meet such demand, and the development of strategies for the deployment of resources, in particular for accommodating temporary changes in demand levels. Capacity planning will involve the examination of alternative strategies. For example, it will be necessary to examine alternative methods for meeting demand level fluctuations. Can and should the amount of resources (e.g. labour) be varied? Is it desirable to maintain a steady level of activity and if so how can this be achieved? Can customers be expected or required to wait for their goods or services? What is the role of inventories and—the question posed above—should the system attempt to meet all potential demand?

Capacity management is rarely simply a matter of determining and providing a required fixed capacity for a system. In most cases there will also be a need to determine whether and how capacity might be adjusted to accommodate changes in demand. The latter is often the more complex problem. Both are aspects of *capacity planning* which will be the focus of our discussion in the following sections.

The shorter term aspects of capacity management, involving the manipulation and deployment of system resources, are achieved through inventory management and activity scheduling. The nature of these two activities is therefore influenced by capacity planning decisions.

CAPACITY PLANNING

The capacity of a system is a reflection of the amount of resources available for the performance of the function. A system has zero capacity unless it can draw upon at least some of each type of resource necessary for the performance of its function. Thus in a transport system the existence of vehicles is insufficient for the creation of capacity. A capacity to transport exists only if vehicles, drivers, consumable materials, etc, are all available. An objective of capacity planning is therefore the determination of the quantity of each resource required. Notice that some resources (e.g. labour and machines) may be stocked, used and reused, whilst others (e.g. materials) will be acquired, i.e. input, and consumed. In the former case a decision must be made as to the amount of each resource to have in stock, whilst in the latter the input rate must be determined. In many cases such decisions will reflect assessment of the *average* demand rate expected over a period of time. If resources are available to satisfy average demand, then if fluctuations about this level can be accommodated, a satisfactory capacity situation exists. In some cases, periods during which demand exceeds average may be offset against periods below average. In others, demand above expectations may be lost, hence there may be some justification in providing capacity in excess of expected requirements as a form of safeguard. In most cases demand levels lower than expected will give rise to either an under-utilization of capacity and/or a build up of resources beyond expectations. Failure to satisfy either resource productivity and/or customer service criteria can therefore result from inaccurate assessment of average demand levels, and hence the provision of too little or too much capacity, or failure to provide for adequate adjustment to fluctuations about these levels through either under-estimation of the extent of demand fluctuations or insufficient capacity flexibility. For example, in any system where resources are stocked an overestimate of demand will lead to an underutilization of resources. An underestimate of demand in any system may, sooner or later, affect customer service—sooner in the case of systems without output stocks and later where output stocks exist. In structures SOS and DOS where function is undertaken in anticipation of demand an overestimate of demand will lead to increasing stock. Temporary under or overcapacity resulting from inability to adapt to demand level fluctuations may give rise to precisely the same effects.

Whilst it is convenient to consider capacity planning to occur in two stages, i.e. determination of average levels and planning for meeting variations about this level, these two aspects are clearly interdependent. The capacity provided may be influenced by the manner in which adjustments may be made. Constraints on adjustment particularly limitations on ability to accommodate short term excess demand may necessitate provision of 'excess' capacity.

Strategies for capacity management

Faced with fluctuating and uncertain demand levels there are two basic strategies.

Strategy 1—Provide for efficient adjustment or variation of system capacity

In most situations it will be both possible and desirable to adopt this strategy. Usually system capacity can be changed within certain limits, perhaps not instantaneously but certainly with little delay. Temporarily more useful capacity might be created by providing more resources and/or providing for their more efficient or intense utilization. Temporary reductions in capacity might be achieved through the transfer of resources to other functions or of course the temporary reduction in the resources on hand or the input rate of resources consumed.

In *supply* systems, e.g. supermarkets, such a strategy is employed as the principal means for accommodating inevitably fluctuating demand levels. Consider the supermarket check-out system. In periods of low demand some of the resources, i.e. the staff, can be transferred to other functions such as re-stocking shelves. During periods of high demand staff resources may be temporarily increased by transfer from other functions to provide 'double-manning' i.e. operation of cash register, and wrapping and loading, in turn providing for more intensive utilization of the other resources, i.e. the cash register, counter, etc. Similarly in certain *manufacture* systems capacity released during periods of low demand might be employed on rectification or service work, whilst peak demand periods might be accommodated by temporary increase in resource levels through, for example, overtime working and more intensive use of equipment perhaps through deferral of maintenance work. In some cases capacity might effectively be increased by subcontracting work. To some extent this strategy might be appropriate in the management of *transport* and *service* systems. In both cases maintenance and service work might be scheduled for periods of low demand. Flexible work shift patterns might be employed, overtime working introduced, etc.

Exhibit 7.2 lists some of the means available for the adjustment of system capacity.

Strategy 2—Eliminate or reduce the need for adjustments in system capacity

In some cases it may be impossible, undesirable or time-consuming to provide for temporary adjustment in system capacity. In general it will be difficult to provide for temporary capacity adjustments in systems which employ large quantities of a large variety of resources, without incurring considerable expense and/or delay. Complex process plants which normally work around the clock present little scope for capacity adjustments to meet temporary demand increases, whilst reductions in demand will often give rise to underutilization of major resources. Similarly in systems which utilize highly specialized resources, such as skilled labour, it may be desirable to avoid the need for temporary capacity adjustments.

In such situations a strategy of minimizing the need for system capacity adjustments will be more appealing. The adoption of such a strategy might involve the provision of excess capacity and therefore the acceptance of perhaps considerable underutilization of resources, in order to increase the probability of being able to meet high or even maximum demand. Such an approach might be desirable where there is little possibility of providing temporary increases in capacity, and where customer service is of paramount importance. Examples of such situations might include an emergency ambulance service, power station or hospital emergency ward. The provision of some excess capacity, yet insufficient to meet maximum demand, necessitates the acceptance that during periods of peak demand either customers will be lost, or they must wait or queue until demand levels fall. In practice such an approach is frequently adopted, for in many cases very considerable excess resources must be provided to ensure that peak demand can be satisfied.

In systems where output stocks can exist, the provision of inventories of goods is a conventional strategy for the smoothing of demand. Such inventories not only insulate the function from fluctuations in demand levels and thus facilitate the use of relatively stable resource levels and high utilization, but also enable customers to be provided with goods with little delay. Many systems operate in this fashion, e.g. the production of domestic appliances, vehicles, building materials, etc. The manufacture of Christmas cards, fireworks, and goods subject to seasonal demand fluctuations is often undertaken on this basis, especially where

MEANS FOR CAPACITY ADJUSTMENT

RESOURCES	CAPACITY INCREASES	CAPACITY REDUCTIONS
ALL	Subcontract some work Buy rather than make (manufacture only)	Retrieve some previously subcontracted work Make rather than buy (manufacture only)
CONSUMED Material	Reduce material content Substitute more readily available materials Increase supply schedules Transfer from other jobs	Reduce supply schedules Transfer to other jobs

FIXED
Machines
	Scheduling of activities i.e. speed and load increases
	Scheduling of maintenance i.e. defer, hire or transfer from other functions
	Hours worked i.e. overtime
	    rearrangement of hours, shifts,
	    holidays.
		Scheduling of maintenance i.e. advance
		Subcontract or transfer to other functions
		Hours worked i.e. short time rearrangement of hours, shifts, holidays.

Labour	Workforce size i.e. manning levels
	    temporary labour
	    transfer from other areas
		Workforce size i.e. layoffs, transfer to other areas.

Exhibit 7.2 Means available for capacity adjustment

resources cannot readily be utilized for other purposes. A similar situation exists where customer waiting or queueing is feasible. In such cases despite a fluctuating demand rate, the rate at which customers are dealt with, i.e. the system capacity, might remain fairly stable. Bus and rail services are frequently intended to operate in this manner. Similar situations might exist at times of peak demand in both manufacture and supply, e.g. the bespoke tailor, the retail shop, etc.

This strategy for capacity management is summarized in Exhibit 7.3, from which it will be seen that there are four basic approaches which might be adopted individually or in combination, i.e.:

(1) Maintain 'excess' capacity (and reduced resource productivity).
(2) Accept loss of customers.
(3) Require customer queueing or waiting (reduced customer service).
(4) Provide output stocks.

| | RELEVANCE | |
APPROACH	Accommodation of temporary demand increases	Accommodation of temporary demand reductions
(1) Maintain 'excess' capacity	Yes	Not directly relevant
(2) Accept loss of customers	Yes—in effect some demand ignored	Not directly relevant
(3) Customer queuing or waiting	Yes—queue increases	Yes—queue reduces
(4) Output stocks	Yes—stocks reduce	Yes—stocks increase

Exhibit 7.3 Means for eliminating or reducing the need for adjustments in system capacity

Notice that only approaches (3) and (4) permit the accommodation of temporary demand reduction without the risk of reductions in capacity utilization. Notice also that the provision of excess capacity alone is rarely a sufficient basis for accommodating demand fluctuations. In most cases it will be necessary to take some action aimed at reducing or smoothing the *effect* of fluctuations on the function, i.e. approaches (2), (3) and (4) above.

Capacity strategy and system structure

In examining internal factors influencing the structure of operating systems we identified the importance of capacity or balance. The discussion above confirms this observation. Decisions taken in capacity management, particularly the decision to employ the strategy of minimizing the need for capacity adjustment, will often have implications for structure. Given the feasibility of stocks, their use will depend largely upon capacity planning considerations. Since the provision of stocks is one of the principal approaches for dealing with demand fluctuations, the feasibility of their use will have a major impact on the capacity problem. Capacity planning decisions particularly in the choice of strategy for dealing with demand fluctuations, relate to the *choice* of system structure and vice versa, whilst *changes* in structure may result from failures in capacity planning (i.e. provision of inappropriate capacity) or be a feature of the strategy adopted.

In many cases organizations would prefer to have demand level fluctuations eliminated or reduced. To some extent they may be able to smooth demand by offering inducements or by requiring customers to wait. Failing or following these efforts to reduce the effects of remaining demand, fluctuations will be preferred, i.e. strategy (2), and finally operating systems will seek to accommodate fluctuations by adjusting capacity. However, the opportunities (or indeed the need) to adopt these strategies for capacity planning will be influenced or limited by the structure of the system. Initially we need only consider those structures in which resource stocks are maintained, since capacity planning as outlined above is needed only where resources are acquired in anticipation of requirements. In these cases capacity planning will aim to deal with uncertainty on the number of orders to be received and perhaps also uncertainty on the resource needed to satisfy the orders received. The feasibility of each strategy for each of the four structures is outlined in Exhibit 7.4 and discussed below.

Structure SOS, with function in anticipation of demand, permits accommodation of fluctuations in demand level through the use of physical stocks, which not only protect the function against unexpected changes in demand level, but also permit a relatively stable level of function activity and thus high capacity utilization. The stock levels employed will often reflect the variability of demand and the 'service level' to be provided, i.e. the acceptable level of probability of stock-out situations with the consequent risk of loss of trade or customer waiting.

Since systems with structure SOD will in most cases have relatively fixed capacity, during temporary high demand periods they will require either customer waiting or suffer loss of trade. Since some excess capacity will normally be provided, capacity utilization will often be low, especially when demand is highly variable.

Since structures SCO and SQO do not permit function in anticipation of demand, either a relatively fixed capacity will be underutilized (SCO) despite efforts to maximize ability of system to adjust, or customer queueing will be required. The queue size will depend upon relative levels and variabilities of demand and function capacity, and in some cases through the use of scheduling (e.g. appointment) systems, customer queueing can be planned.

The relative value of strategy (1), provide for efficient adjustment of system capacity, and strategy (2), eliminate or reduce need for adjustment in system capacity, is influenced by system structure feasibility. Other factors will, how-

STRATEGY / STRUCTURE	1. PROVIDE FOR EFFICIENT ADJUSTMENT OF SYSTEM CAPACITY	2. ELIMINATE OR REDUCE NEED FOR ADJUSTMENT IN SYSTEMS			
		a. Maintain 'excess' capacity	b. Reduce or smooth effect of demand level fluctuations		
			I. Fix upper capacity limit with effect of		II. Use stock to absorb demand fluctuations
			Loss of trade	Customer queueing/waiting ** *	
SOS 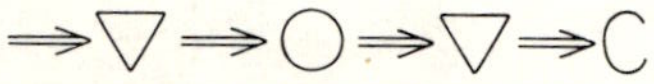	Feasible and often desirable to supplement strategy 2bII	Feasible, but not necessary	Feasible, but not normally necessary	Waiting feasible but not normally necessary	Feasible, and normally adopted
SOD 	Feasible and often desirable to supplement strategy 2bI	Feasible, and may be necessary in conjunction with or instead of 2bI	Feasible and normally adopted	Waiting feasible and normally adopted	Not feasible
SCO 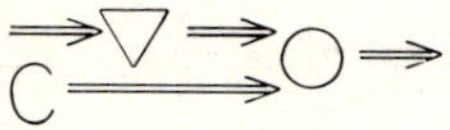	Feasible, and desirable in conjunction with 2a	Feasible and normally adopted	Might be feasible depending on nature of function	Not feasible	Not feasible
SQO 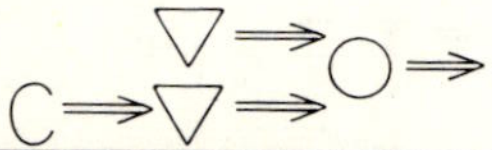	Feasible and often desirable to supplement strategy 2bI	Feasible but not necessary	Feasible and might be adopted with	Queueing feasible and normally adopted	Not feasible

* Customer push
** Customer pull

Exhibit 7.4 Capacity planning strategies (system structures SOS, SOD, SCO, SQO)

ever, influence the choice of strategy. If, for example, there is a limit to the size of output stocks then although strategy 2bII is feasible it may not be possible to rely upon this means for meeting demand level changes. In most situations it will be desirable to consider providing effective capacity adjustment to meet demand level change, but in most cases effective capacity management will depend also upon a preventative strategy, either through the absorption of fluctuation through stock (SOS), or through customer queueing (SQO), and waiting (SOD), whilst in structure SCO low capacity utilization is probably unavoidable. Systems which permit function in anticipation of demand will normally use output stock to protect against demand level fluctuations. Hence the management of finished goods inventory is of crucial importance. Other systems will normally rely upon customer queueing and will, where possible, seek to schedule customer arrivals. In system SOD there will be relatively high capacity and low capacity utilization. Notice that in many situations a mixed strategy will be employed.

Now let us consider structures DOS, DOD and DQO, in which capacity is not maintained in anticipation of demand. Systems with structure DOS will probably utilize resources when they become available. Structures DOD and DQO normally provide for the acquisition of all necessary resources on receipt of customer order. Neither structures DOD or DQO require a forecast of the number of customer orders which will be received since demand can be measured. However, a similar type of forecasting problem *may* occur since the nature of orders will vary and there will often be some uncertainty as to the exact amount of resource required for satisfying these orders which are received. It may therefore be necessary to anticipate capacity adjustment and/or try to avoid it. Exhibit 7.5 indicates the type of strategy which might be employed.

Demand will be forecast for structure DOS in order that intermittent production for stock might be sufficient to satisfy customer needs. The use of output inventories is normally an essential part of the system since, whilst the function is intermittent, demand may be continuous. Inventories will accommodate demand fluctuations hence there will be a need to conserve more or less resources only in the event of significant changes in demand levels (see Exhibit 7.5).

CAPACITY CONTROL

The operation or control elements of capacity management relate to the manipulation and deployment of the resources available to the system. This will be achieved primarily through activity scheduling and inventory management. Capacity planning will influence the nature of inventory, and the management of these inventories will help achieve capacity control. More directly the scheduling of activities will influence the manner in which resources are deployed. Exhibit 7.6 summarizes.

Since capacity control problems derive from the need to meet demand fluctuations, the nature of this problem will be determined by the nature of the

STRATEGY / STRUCTURE	1. PROVIDE FOR EFFICIENT ADJUSTMENT OF SYSTEM CAPACITY	2. ELIMINATE OR REDUCE NEED FOR ADJUSTMENT IN SYSTEMS CAPACITY			
		a. Acquire capacity in excess of expected requirement	b. Reduce or smooth effect of demand level fluctuations		
			I. Fix upper capacity limit with effect of		II. Use stock to absorb demand fluctuations
			Loss of trade	Customer queue-ing/waiting **	
DOS	Sometimes desirable in order to increase or reduce capacity to match demand level changes	Not relevant	Not relevant	Not relevant	Feasible, and normally necessary
DOD	Sometimes necessary with strategy 2a because of uncertain capacity requirements of given customer orders	Sometimes necessary	Not relevant +	Not relevant +	Not feasible
DQO	Sometimes necessary with strategy 2a because of uncertain capacity requirements of given customer orders	Sometimes necessary	Not relevant +	Not relevant +	Not feasible

+ Not relevant since capacity is acquired against given customer orders

* Customer push
** Customer pull

Exhibit 7.5 Possible capacity management implications (system structures **DOS, DOD, DQO**)

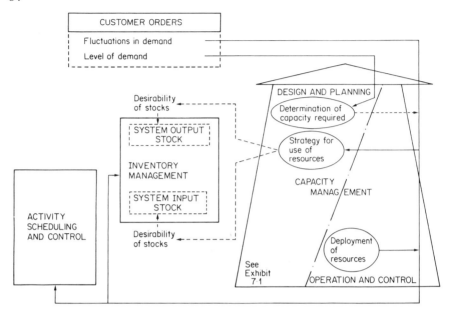

Exhibit 7.6 Capacity control

strategy adopted, i.e. the relative importance attached to the need to adjust capacity and the avoidance of such adjustments. In particular activity scheduling will normally be more complex, and of course more important if the strategy adopted emphasizes the need to provide for efficient adjustment of capacity levels (strategy 1). Thus in our review of the means for capacity adjustment scheduling actions are evident. A strategy based more upon the smoothing of demand level fluctuations will place emphasis upon inventory management.

EXAMPLES

The nature of the capacity management problem for three of the systems listed in Chapter 3 are outlined below and summarized in Exhibit 7.7.

Fire service

In this case we can consider the normal system to be structure SCO, since it will usually be the intention of the fire service to ensure that appliances, etc, are 'standing-by' for calls. There will of course be situations, such as major disasters, in which all available resources are committed and hence new customers must wait. In such cases the system structure is, in effect, DQO. Here we shall consider the capacity management problem for the normal, i.e. the intended case (structure SCO).

Expected demand will be forecast, and capacity will doubtless be provided sufficient to meet substantially more than average demand. In effect therefore

EXAMPLE	NORMAL SYSTEM STRUCTURE	DETERMINATION of CAPACITY REQUIRED	CAPACITY PLANNING STRATEGY	PRINCIPAL OBJECTIVES
Builder (of 'one-off' houses)	DOD	Demand is measured. Capacity is provided to meet each demand	1 and/or 2a i.e. Provide for capacity adjustment and/or excess capacity	Maximum customer service (in particular through minimizing completion time) + Maximum utilization of (consumed) resource
Fire Service	SCO	Expected demand is forecast. Capacity is provided to meet maximum or near maximum demand	2a with some possibility of 1. i.e. Eliminate or reduce need for adjustment in system by providing excess capacity with further possibility of providing some capacity adjustment	No customer queueing + High resource productivity
Furniture Removal	SQO	Expected demand is forecast. Capacity is provided to meet average or 'sufficient' demand	2bI (with customer queueing and possibly loss of trade), with possibility of 1. i.e. Eliminate or reduce need for adjustment in system capacity by smoothing effect of demand level fluctuations through fixing upper capacity limit and accepting loss of trade with the further possibility of providing some capacity adjustment	Minimum customer queueing and/or loss of trade + High resource productivity

Exhibit 7.7 The nature of capacity management—examples

capacity planning strategy 2, i.e. 'eliminate or reduce need for adjustment in system capacity' through the 'provision of excess capacity' is adopted. Undoubtedly the emphasis will be upon this approach.

However, there may also be some possibility of providing for some degree of capacity adjustment (i.e. strategy 1) through, for example, subcontracting work (to neighbouring fire services), transferring resources from other jobs (e.g. maintenance, practices, etc) and temporary yet rapid increases in the labour force or hours worked (use of standby labour etc).

The principal objective in capacity planning will be the maintenance of the system structure, i.e. to ensure that no customer queueing occurs, coupled of course with reasonably high capacity utilization. In practice the customer service criterion will often be satisfactorily attained, at the cost of poor resource productivity.

Furniture Removal

This differs from the above example in offering a far wider range of alternative approaches to capacity management. Here structure SQO will normally apply since resources will be stocked and will be utilized on jobs drawn from an available queue.

Again expected demand will be forecast and on this basis a certain capacity will be provided. Unlike the fire service, there will be no necessity to meet high demand and in practice capacity will be provided sufficient to meet near average demand or 'sufficient' demand—sufficient being defined by management as that proportion of available demand which management choose to seek to satisfy. Having established a normal capacity in this way, the operations manager must choose whether, and if so how, to provide for efficient capacity adjustment (strategy 1) and/or whether, and if so how, to eliminate or reduce the need for such adjustments (strategy 2). Both strategies are feasible and might be implemented as follows:

(1) Provision for efficient adjustment of system capacity, through for example subcontracting work, transfer of resources from other jobs, rescheduling activities, temporary changes in manning or hours.
(2) Eliminate or reduce need for adjustment in system capacity through:

(a) Providing excess capacity (although this is neither normally necessary nor adopted).
(b) Reducing or smoothing the effect of demand level charges by I fixing an upper capacity limit and normally requiring customers to queue, and in some cases accepting some loss of trade.

In all cases the objective in capacity management will be the provision of an acceptable level of customer service by minimizing average waiting time or queue length and thus minimizing the loss of customers, coupled with providing for high resource productivity, i.e. utilization of resources.

Builder

In this case we can consider the case of the special 'one-off' house builder who neither has a permanent labour force nor owns much equipment. In effect on receipt of an order from a customer he acquires all the resources necessary for the satisfaction of the order, thus the system is essentially structure DOD. In this case the capacity management problem is relatively simple. The determination of capacity needs is a straightforward problem, since no capacity (i.e. no resources) are provided prior to receipt of the customer's order. There is therefore no need to attempt to forecast demand. Demand is in effect measured and sufficient capacity is then provided to meet that demand, i.e. to satisfy that order. Theoretically at least, the development of a strategy for the utilization of resources is straightforward, since there is never the need to provide for efficient adjustment of capacity nor the need to eliminate or reduce the need for such adjustment. In practice, however, since in such situations all orders will be different, it may be difficult to determine exactly how much of each resource is required. For example the quantity of each type of building material can only be estimated, the amount of labour required will depend upon a variety of factors, etc. In practice therefore there will in effect be a need to *either* ensure that further capacity can be provided if insufficient is initially obtained *and/or* to eliminate or reduce the need for adjustment in system capacity by the provision of excess capacity. Hence the capacity planning strategy adopted will often consist of:

(1) Provision for efficient adjustment of system capacity, through, for example, subcontracting, transferring material, labour and machinery from other jobs, increasing hours worked and/or labour force, etc.

and/or

(2) Eliminate or reduce need for adjustment in system capacity by providing excess capacity.

Customer service will be one principal objective in determining and managing capacity in such systems. Apart from price, which will in part be influenced by the resource utilization expected, delivery or completion time will be the main variable affecting the probability of receiving orders. The customer service objective will therefore be manifest in attempts to minimize building time. In addition the building operations manager will of course seek to ensure that his resource productivity is high, in order that his costs can be kept down. He will therefore seek to satisfy a partly competing objective by ensuring that equipment, labour and materials are all sufficiently highly utilized. He may therefore be inclined to favour an emphasis on strategy 1 above in planning his capacity requirements.

SUMMARY OF CHAPTER 7

The effective management of system capacity is the *most important responsibility* of operations management.

The *objective of capacity management* is to match or balance the level of operations to the level of demand. *Capacity planning*—the more important aspect of capacity management—involves the study of likely medium to long-term demand patterns, *the determination of the capacity required* to meet such demand and the *development of strategies for the deployment of resources*, in particular for accommodating temporary changes in demand levels.

There are *two basic strategies* for capacity planning, i.e.

(1) Provide for efficient adjustment or variation of system capacity.
and
(2) Eliminate or reduce the need for adjustments in system capacity.

Adjustment of system capacity might be achieved through, for example, sub-contracting work (or vice versa), rescheduling activities, changes in manning levels or hours, etc.

Elimination or reduction of need to adjust capacity might be achieved by providing excess capacity, accepting loss of customers, requiring customer queueing and providing output stocks.

Decisions taken in capacity planning, particularly the decision to employ the strategy of minimizing the need for capacity adjustment will often have *implications for system structures*.

The opportunities (and need) to adopt these strategies will be influenced or linked by system structure feasibility.

Capacity control is achieved largely through inventory management, and activity scheduling decisions.

CHAPTER 8

The Nature of Activity Scheduling

Activity scheduling problems are closely related to those of capacity and inventory management, all three being influenced by, and an influence on, system structure. The nature of activity scheduling was outlined in Chapter 6 where it was established that information input to scheduling decisions, derived from either the customer or from system output stocks, whilst information was output to either system input stock or the supplier. Clearly the structure of the system, in that this will reflect the presence or absence of physical stock, is related to the nature of the activity scheduling problem. In demand pull structures where no output stocks exist, activity scheduling decisions will be directly influenced by customer demand, whilst the absence of input stocks will necessitate a direct influence of activity scheduling over suppliers.

This relationship of activity scheduling with inventory management was also noted in Chapter 6. Inventory management will be shown to be concerned with the manipulation of flow in parts of the system. Activity scheduling is also concerned with the manipulation of flow, indeed inventory management decisions with decisions in activity scheduling together influence flow throughout the entire system. Decisions in both areas relate to the timing or scheduling of activities or events. Both inventory and activity scheduling can therefore be seen as aspects of flow scheduling in systems. Both may be seen as short to medium-term problem areas.

Exhibit 8.1 indicates the manner in which customers demand is accommodated throughout the system by a combination of scheduling decisions, i.e. inventory and activity scheduling. Consider, for example, system structure SOS—'function from stock, to stock, to customer'. Here the customer will directly influence final inventory output flow, inventory management will influence inventory input in order to meet output requirements, activity scheduling will influence function input to satisfy inventory input requirements, etc. Similar situations apply in the case of demand push systems. In structure SQO, for example, customers' demands will influence the management of customer inventories (queues), which will in turn influence activity scheduling, finally influencing the management of input resource stocks. Each system therefore is 'covered' by a 'chain' of operations scheduling decisions, which in effect transfer or convert customers' direct demands on the system to demand on resource suppliers. Activity scheduling is a link in this chain. The structure of the system will influence the location

90

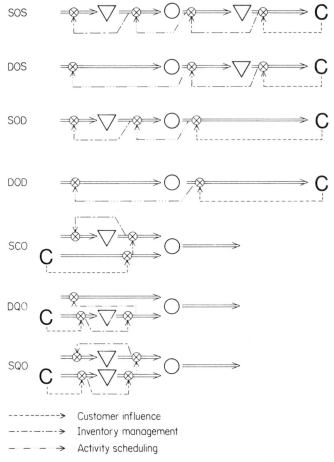

SOS

DOS

SOD

DOD

SCO

DQO

SQO

- - - - - - - - ->   Customer influence
- - · - - · - ->   Inventory management
- — — — — ->   Activity scheduling

Exhibit 8.1 Operations scheduling relationships

of this link, i.e. the relationship of activity scheduling with inventory management and/or the customer and/or the supplier.

CUSTOMER INFLUENCE

Activity scheduling together with inventory management *accommodates* customer demand throughout the system. Customers will have some direct influence over part of the system activities, the remainder being influenced by decisions of inventory management and activity scheduling. The various operating system structures provide for different degrees of customer influence on the system. For example in structure SOS (function from stock, to stock, to customer) the customer directly influences output inventory, i.e. the customer is supplied from stock and the stock policy is influenced by the characteristics of expected demand. Stock policy in turn influences the function, etc. A similar situation applies in structure DOS and also in structure DQO and SQO. In the latter cases where customer demand push applies, customers again influence the function through stock, in this case customer input queues. As with output stock, excepting through the manipulation of demand, the only means for the control of queue levels is through function capacity adjustment. In structure SOD and DOD the customer directly influences the system function, likewise structure SCO.

One important aspect of operations scheduling therefore relates to the accommodation of direct customer influences. This is the *external* aspect of the scheduling problem, i.e. satisfying customer requirements in the form of required delivery or service dates. This aspect of scheduling is largely related to the criteria of customer service whilst *internal* considerations are largely related to the productivity criterion.

ORIENTATION OF ACTIVITY SCHEDULING

In each basic system structure the customer will to some extent have an influence upon when things happen. The customer will request or otherwise indicate, e.g. by his arrival, attention at a specific time—the *due date*. To some extent the management of the operating system will be directly influenced by such external requirements, the due date requirement being accommodated directly through inventory management *or* activity scheduling of the function. In the context of activity scheduling the external due date problem is of particular importance in structures SOD, DOD and SCO.

Internally oriented activity scheduling is of particular importance in structures SOS, DOS, DQO and SQO, indeed the possibility of adopting internally-oriented activity scheduling in such situations often gives rise to relatively high resource utilization compared to other structures. Some structures (SOS, DOS and SQO) tend to consist of repetitive functions (i.e. function predictability), hence the activity scheduling procedures employed are those normally associated with repetitive operations. Structure DQO is an exception in being non-repetitive.

Exhibit 8.2 indicates the purpose of activity scheduling for each of the seven basic structures. The exhibit suggests the type of approach adopted in satisfying both external and internally-oriented requirements in each of the structures. Since externally-oriented scheduling is concerned primarily with the accommodation of due date or delivery criteria, scheduling decisions which take direct account of customer needs normally require a 'due date' approach, whilst internally-oriented scheduling may involve a variety of techniques and procedures, in all cases the objective being related to the provision of satisfactory resource productivity.

This is perhaps an appropriate point at which to remind ourselves that the actions of operations management will be influenced by their perception of the system, or by the normal or intended nature of the systems (Chapter 5). For example, if a system is intended or believed to have structure SOS, operations

Structure / Requirement of Activity Scheduling	External requirements, i.e. to satisfy due date and customer service criteria	Internal requirements, i.e. to satisfy flow and resource productivity criteria
SOS	(Accommodated by output inventory decisions)	Accommodated through activity scheduling for function
DOS	(Accommodated by output inventory decisions)	Accommodated through activity scheduling for function
SOD	Accommodated by due date scheduling of function	Accommodated by scheduling for function within constraint imposed by external requirements
DOD	Accommodated by due date scheduling of function and resources input	Accommodated by scheduling for function within constraint imposed by external requirements
SCO	Accommodated by due date scheduling of function	Accommodated by scheduling for function within constraint imposed by external requirements
DQO	(Accommodated by customer queue/inventory decisions)	Accommodated through activity scheduling for function
SQO	(Accommodated by customer queue/inventory decisions)	Accommodated through activity scheduling for function

Exhibit 8.2 The purpose of activity scheduling

management will schedule activities without direct reference to customer due dates. They will schedule operations to provide output to stock. If, however, output stock has fallen to zero, perhaps because of an increase in demand, structure SOD will in fact exist yet an externally oriented scheduling strategy may not be employed—at least not immediately.

For this reason in discussing the influence of system structure on operations management strategies we must consider perceived, intended or normal structures.

The nature of due date scheduling, an essential aspect of externally oriented activity scheduling is discussed below. The nature and objectives of internally-oriented activity scheduling are also described. The procedures available for activity scheduling are discussed in Chapter 11, whilst the role of inventory management in the scheduling context is examined in Chapter 9.

Externally-oriented scheduling strategy

Scheduling primarily for the accommodation of external (i.e. customer) requirements will normally involve due date considerations. Due date scheduling will usually consist essentially of a reverse scheduling procedure whereby successive lead times (plus allowances) are subtracted from a required due or delivery date, in order to obtain required starting dates for activities; Exhibit 8.3 illustrates in Gantt chart form. This treatment assumes:

(1) That the due dates are feasible, and
(2) 'Simple' functions.

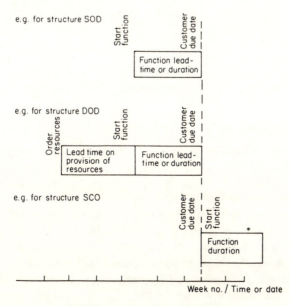

Exhibit 8.3 Due date scheduling

Whilst such conditions might apply in some situations (e.g. in supply, the function is simple and can realistically be expressed in scheduling as a simple activity, usually with insignificant duration), in many cases they do not. For example, a function (e.g. manufacture or service) may consist of multiple activities, and furthermore their normal duration may not permit the satisfaction of a required due date. For either such reasons due date scheduling may be complex and require a more sophisticated approach. (See Chapter 11.)

Internally-oriented scheduling strategy

Such scheduling normally takes place within the constraints, etc, determined by external considerations, and therefore relates essentially to the function rather than the system as a whole. Often the major considerations will be efficient flow capacity utilization, coordination of activities (e.g. timing) and in all cases the objective will be the maximization of function productivity.

The dispatching problem will be evident in many situations, i.e. 'which of the available and feasible jobs to do next?' Dispatching problems will normally apply in 'from stock' situations where due date considerations do not intervene (e.g. systems SOS and SQO). The need for the coordination of activities will often be evident, especially where parallel activities occur, and in such cases a Network Analysis approach may be relevant. In system structures DOS and DQO a batching situation may apply, whilst in systems DQO and SQO the function may be timetabled, i.e. may be scheduled to occur regularly in order to meet a forecast demand. Notice that, as the emphasis is upon the function similar problems occur in both demand pull and push situations.

EXAMPLES

The nature of the activity scheduling problem for three of the systems listed in Chapter 3 are outlined below and summarized in Exhibit 8.4.

Bespoke tailor (structure SOD)

In this case an externally-oriented activity scheduling strategy will be employed. The customers will directly influence the entire system, excluding input resource stocks. The chain of scheduling decisions between customer and suppliers will be provided by activity scheduling and input inventory management. Customer orders, together with their required, quoted and/or agreed delivery dates will be input directly to the activity scheduling function along with necessary information on the work content, method of manufacture and resource requirements. A reverse scheduling procedure will normally be applied, all separate operations being scheduled to commence in order that, given available capacity limits, the garment will be completed on or before the required due date. For simple garments this reverse scheduling procedure might involve one

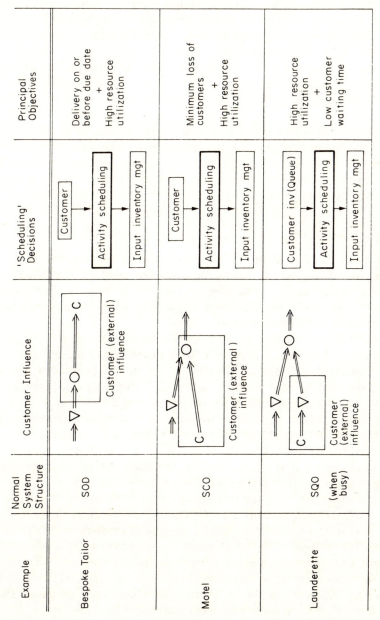

Exhibit 8.4 The nature of activity scheduling and control—examples

operation only, i.e. 'manufacture garment', the normal throughput time for manufacture of such a garment being known with some accuracy. For more complicated garments, especially those requiring several different operations using different resources, the reverse scheduling procedure may recognize several separate operations, each with known or estimated throughput time, and each scheduled to be performed sequentially or, in some cases, concurrently (e.g. concurrent manufacture of the pieces of a suit).

Materials will be withdrawn from input resource inventories according to the specified manufacturing schedule. Inventories of consumable items will therefore be depleted which will in turn lead to the placing of replenishment orders.

The principal objective of activity scheduling in this case will be the provision of high due date performance, i.e. the provision of a high proportion of goods on or before due date *or* the minimization of lateness, e.g. average lateness of jobs or the percentage of late jobs. A simultaneous objective will be the achievement of high resource utilization.

Motel (structure SCO)

Here, much like the situation of the bespoke tailor the customer directly influences the entire system excluding input resource stocks. For this reason an externally-oriented activity scheduling strategy is appropriate and the chain of scheduling decisions between customer and supplier comprises activity scheduling and input inventory management.

In this case, the customer due date will be his time of arrival, whether this follows a prior booking or, as is more likely in the case of a motel (but not a hotel), 'unannounced'. The management of the motel operation will seek to accommodate the customer from this due date. Failure to provide facilities will normally result in loss of custom, for the customer will not normally wait until present residents vacate rooms. (If customers, for whatever reason, are required and are able to wait, the structure may change to SQO, and a different scheduling situation will apply—as in the case of the launderette, see below.) The customer therefore requests a service from or when he arrives, the service or processing time through the function, being either known, having been specified in advance or at the due date, or uncertain (but with perhaps a known minimum duration, i.e. one day). His arrival leads immediately to the depletion of resource input stocks, in that a room will be occupied. Consumable resource stocks will also be depleted. The replenishment of consumable resource stocks will be influenced by inventory management procedures, whilst further rooms, etc, will only be added to resource stocks if total capacity is to be increased. Thus the activity scheduling problem in this case is relatively straightforward, customers simply being allocated on a first come/first served basis to available resources.

In normal motel situations, i.e. no prior booking by customers, decisions on activity scheduling will not significantly affect either the customer service or the resource productivity criterion. The management of the motel will wish to en sure high resource (i.e. room) occupancy, and at the same time try to ensure that the minimum number of customers are turned away or lost. Capacity management decisions will however be of prime importance in both cases. In the case of some or all customers being expected and able to specify required due dates prior to their arrival, i.e. to book their accommodation in advance, it may be possible through activity scheduling to affect both occupancy and service. For example, some customers (requiring for example short stays at normally busy times) may be turned away in order that others (requiring, for example, longer stays) may be accommodated. Additionally, or alternatively, some customers may be offered accommodation at other times, i.e. they may be offered other due dates. In such cases some due date performance is forfeited for increased resource utilization, in much the same manner as will often occur in customer pull structures such as the bespoke tailor (see above).

Launderette

Whilst it might not always operate in this manner, a launderette provides for customer queueing. Customers who cannot be served immediately on arrival will not normally be lost unless the queue length and/or the expected waiting time is substantial. When customer queues occur the system operates with structure SQO, otherwise structure SCO will apply. We shall consider the structure SQO situation in which the direct external influence extends only to the customer queue. Thus the chain of scheduling decisions is provided through management of customer inventory (i.e. the queue), activity scheduling and resource inventory management. The activity scheduling problem in this case requires an internally-oriented strategy, all considerations of customer service being accommodated by inventory and capacity management. The principal objective of activity scheduling will therefore be the achievement of high resource productivity, i.e. high facility utilization.

The customer's time of arrival is his or her due date, hence because of the queueing situation due date performance will be poor. For most customers the time at which service commences will be later than the due date. At this time resource inventories are depleted. Consumable items are used which in turn will effect replenishment decisions through inventory management, however as with the motel, non-consumable resources will not be replenished expecting to increase capacity, or for renewal.

In this situation activity scheduling is relatively trivial. Processing times are largely fixed. Furthermore it is, for example, unlikely that each customer can be scheduled through the function since in most cases customers will be served on a first come/first served basis through those facilities which first became available to them. In fact, therefore, the customers will schedule themselves through the facilities, however some activity scheduling decisions can be made since it will often be possible to establish activity scheduling rules which customers will follow. For example, in a two stage system, i.e. washer and dryer, customers who have used washers may be given priority on dryers.

98

SUMMARY OF CHAPTER 8

The activity scheduling and control problem is closely related to that of inventory management, and both relate to the problem of capacity management in particular the control of capacity.

Activity scheduling is concerned with the manipulation of *flow in the system*, and the *timing of events and activities*.

Activity scheduling decisions with inventory management decisions *accommodate customer demand throughout the system*.

An *externally*-oriented scheduling strategy is concerned directly with the satisfaction of customer requirement in the form of *due dates*, and is concerned largely with the customer service criterion.

An *internally*-oriented scheduling strategy relates essentially to the function, and is concerned largely with the resource productivity criterion.

CHAPTER 9

The Nature of Inventory Management

In Chapter 8 we identified the role of inventory management as part of the chain of scheduling decisions between customers and suppliers. We noted that both activity scheduling and inventory management decisions would effect flows through the system. In this chapter we shall examine the nature of the inventory management problem within this broader scheduling and flow context.

At our level of analysis inventories can exist at two stages in the system, and can consist of two 'types' of items. The existence of inventories of *output goods* will be determined by feasibility and desirability considerations, the latter being influenced by capacity planning considerations. The items in output stock will come from the function and pass to the customer. *Input stocks* may consist of *consumable items* (e.g. materials) which will eventually be converted, transferred or moved by the function and/or *non-consumable items* (e.g. labour and machines) utilized in the performance of the function. The provision of stocks of such items will also be a function of feasibility and desirability considerations, the latter being influenced by capacity planning decisions.

The management of input inventories is directly related to capacity management, whilst the management of output inventories is indirectly concerned with the control of capacity, since the provision of such inventories will influence the manner in which resources are provided and utilized.

From a capacity viewpoint, output inventories provide a form of damper, insulating the function from fluctuations in demand level. The management of output inventories is concerned with the regulation of the flow of items. The management of input stocks of consumed items is much the same type of flow control problem. Output stocks exist to accommodate short-term differences in input and output rates, i.e. the rate of provision of items into stock from the function and the rate of depletion of stock to the customer. Similarly input stocks of consumed items exist to accommodate short term differences in the supply and consumption rate. The extent to which the level and composition of stocks vary is therefore a measure of their usefulness. Lack of variation might suggest the lack of need for stock.

The management of non-consumed input stocks constitutes a somewhat different type of problem. Here the emphasis is upon acquisition/replacement and maintenance rather than repeated acquisition.

Exhibit 9.1. summarizes the above points. The acquisition of resources which

		STAGE IN SYSTEM	
		System Input Stocks	System Output Stocks
TYPE OF ITEMS	Consumed	Control of system capacity Accommodates short term difference in supply and consumption rates Control of flow of resources and customer	Implementation of aspect of capacity management strategy Accommodates short term differences in function and demand rates Control of flow of goods
	Non-consumed	Planning of system capacity Acquisition and maintenance	

Exhibit 9.1 The nature of inventory management

are not consumed and therefore remain to be reused presents inventory planning but not inventory control type problems. In such cases there is the need to determine the level or quantity of resources required and the timing of their replacement but not the need to provide for their continual replenishment. In general inventory control focusses on the replenishment problem, which does not occur, at least not to the same extent, for non-consumed items. There may, of course, be the need to replace and maintain labour and equipment, and such requirements may be related to the level of demand placed on the system. The problem however is not one of managing flows of resources but largely one of acquisition. The management of inventories of non-consumed items is therefore primarily concerned with capacity planning through the determination and acquisition of the capacity required rather than the use of resources (see Exhibit 7.5).

INVENTORIES OF CONSUMED ITEMS

System output inventory

Structures SOS and DOS require output stocks. Customer demand is satisfied from such stocks which in turn are replenished from the function. The information flows in the opposite direction to the physical flow, hence, in the case of intermittent stock replenishment, customer orders will be received at output inventory, depletion of which will give rise to the dispatch orders to the function.

Stocks may be replenished intermittently or continuously, although in some cases the distinction is more evident in the type of inventory management decisions that are required than in the physical flow into stock.

For our purposes we can consider the nature of customer demand to be given, however a knowledge or estimate of the nature of demand, in particular the demand level and fluctuations, will influence inventory management. The stock levels maintained, and/or the amount of buffer (or safety) stock provided will reflect expected demand levels and fluctuation.

The complexity of the inventory management problem, and the likely effectiveness of inventory management depends upon the variability or unpredictability of stock input and output levels and also upon the opportunities for, and the extent of, control. Thus in certain systems inventory management is likely to be more effective than in others simply because there exists the opportunity for the exercise of closer control.

Exhibit 9.2 shows the two extreme types of situation which may exist in the management of output inventories. (The diagrams do not indicate the *degree* of predictability of inputs and outputs.) In many cases an intermediate situation will exist in which some incomplete control of stock input is available.

Case 1.   Full control of input to stock

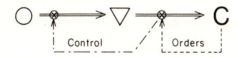

Case 2.   No control of input to stock

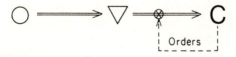

N.B.   In most situations some degree of control of inputs
is possible

Exhibit 9.2 Management of output inventories

Exhibit 9.3 demonstrates the effect of input and output variability on the inventory management problem. The diagrams describe stock replenishment procedures in which inputs are requested when stock falls to a given 'order level', i.e. intermittent replenishment. In each case there exists a need for buffer stock holdings to accommodate variable replenishment lead time usage resulting from either a variable usage rate, i.e. output rate, or a variable input lead time or both. Case A corresponds to a situation in which demand is perfectly predictable (i.e. usage for any given period of time is known), yet where there is incomplete control of input and hence there is some uncertainty on lead times. A

102

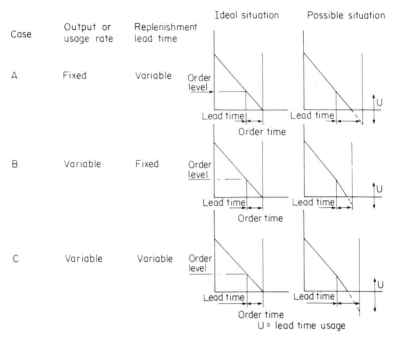

Case	Output or usage rate	Replenishment lead time		Ideal situation	Possible situation

Exhibit 9.3 Inventory management—the effects of variability in input and output

buffer or safety stock must be held lest lead time is greater than anticipated. Precisely the same situation would apply if incomplete control of inputs gave rise to uncertain quantities of input items and of course the problem would be aggravated if there was uncertainty on both replenishment lead times and quantities. Case B represents full control or predictability of input, but unpredictable usage or output. Buffer stocks must be held to protect against unexpectedly high usage during replenishment. Case C combines the worst of both other cases, i.e. uncertain output and input.

Similar situations apply in the case of continuous flow to stock.

The classic stock control problem is that of establishing an inventory policy based upon some control over inputs to satisfy unpredictable output, demand or usage. The situation may resemble, at best, Case B in Exhibit 9.3, i.e. variable demand or usage rate and fixed or predictable input, although in practice the situation may tend towards Case C. As there is no opportunity for control over output a forecast rate must be obtained as a prerequisite for inventory planning and control. Given a forecast output or usage rate per unit time and the variability of that output rate, the following inventory parameters might be established:

(1) For intermittent replenishment;
   (a) Either re-order level or interval, and
   (b) Order quantities

(2) For continous replenishment:
   (a) Input rate, and
   (b) Average stock level required.

The absence of opportunity for significant input control (Case 2, Exhibit 9.2) suggests that there will be little opportunity for the management of inventory. In such cases, there may be the opportunity for determining a preferred buffer or minimum stock level for the inventory which might be established at the commencement of operation of the system. Thus given forecast input and output rates and the variability of these rates, the inventory may be primed, i.e. built up at the initiation of the system. Alternatively, a given inventory capacity may be established, i.e. sufficient space for inventory. In such situations, inadequate inventory will lead to either the starving of the output channel or the blocking of the input channel and because of the absence of opportunities for control, unless input and output are relatively stable, inventory management is likely to be relatively ineffective.

System input inventory (consumed items)

Structures SOS, SOD, SCO and SQO require input resource stocks. The problem of managing the stocks of input resources closely resembles that of system output inventory management. If the function is considered as a customer for input resource stocks then, as with output stocks, demand is satisfied from stock which in turn is replenished from supply. In the case of intermittent inputs, orders will be received into stock through activity schedules. The depletion of stock will give rise to the dispatch of replenishment orders to suppliers.

The activity scheduling function will be responsible for the manner in which input stocks are depleted. Consumable resources will either be scheduled through the function or a throughput rate will have been fixed. In either case we can again consider the nature of demand on stocks to be given; however a knowledge or estimate of the nature of demand, in particular demand level and fluctuations, will again influence input inventory management. As with output stocks, the amount of buffer (or safety) stock provided will reflect expected demand levels and fluctuation as well as the predictability and degree of control available over replenishment. In this case the extremes of control are shown diagrammatically in Exhibit 9.4. The three cases given in Exhibit 9.3 are relevant also in input resource stock management.

Customer input queues

Structures DQO and SQO require input queues (or stocks) of customers. In such systems customers, or items provided by them, are consumed resources. We have argued that in transport and service systems the customer himself, or something provided by him, is an input which differs from other consumed resources only in being beyond the direct control of system management. In

Case 1. Full control of input to stock

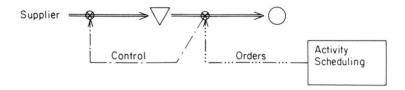

Case 2. No control of input to stock

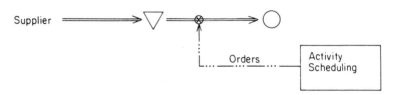

Exhibit 9.4 Management of input resource stocks (consumed items)

other words there is little or no control over the input or arrival of such resources, i.e. their input is unpredictable.

Ostensibly the management of such customer stocks resembles the management of other input consumed resources where little or no control over input to stock exists, e.g. Case 2, Exhibit 9.4. However this case is the reverse of Case 2. There exists no opportunity for control over inputs and thus the inventory is in effect managed through control over the output. Exhibit 9.5 illustrates the two extreme cases. In most situations some output control will be available, hence

Case 1. Full control of output from stock

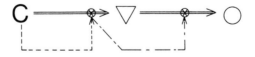

Case 2. No control of output from stock

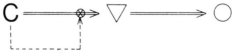

Exhibit 9.5 Management of customer input
stocks

given an estimate of the input rate and the variability of that rate the following inventory parameters might be established.

(1) For intermittent depletion:
    (a) Output intervals or the stock levels at which output is actuated, and
    (b) Output quantities.
(2) For continuous depletion:
    (a) Output rate, and
    (b) Average stock level required.

Inventory management strategies

We have identified three basic inventory control situations. These are summarized in Exhibit 9.6, which also indicates where each may occur in the seven basic operating system structures. The principal distinction is in the location of control, i.e. input or output control. We have also noted the distinction between intermittent and continuous flow. The decisions required for the management of inventory in each situation are summarized in Exhibit 9.7. The relationship of

CASE	DESCRIPTION	EXAMPLES		
			Structure	Inventory
1. control	Input control		SOS	Output
			SOS	Input
			SOD	Input
			SCO	Input
			SQO	Input
2.	No control		DOS	Output
3. control	Output control		DQO	Customer input
			SQO	Customer input

Exhibit 9.6 Basic inventory control situations

		LOCATION OF FLOW CONTROL	
		On Stock Inputs	On Stock Outputs
NATURE OF FLOW	Intermittent	Determine stock replenishment level or interval. Determine input quantity.	Determine stock levels or intervals at which output is to take place. Determine output quantity.
	Continuous	Determine average (or safety) stock levels required. Determine stock replenishment (input) rate.	Determined average stock levels required. Determine stock output rate.

Exhibit 9.7 Inventory management strategies (consumed items)

inventory management and activity scheduling is again evident from the preceding discussion.

Available quantitative models and techniques for inventory management are inadequate as a basis for effective inventory management in many of these three cases. Clearly input and output control models are required, each of which must be capable of dealing with intermittent and continuous replenishment and/or depletion. We shall review available models and techniques in Chapter 12.

EXAMPLES

The nature of the inventory management problem for three of the systems identified in Chapter 3 is discussed below and summarized in Exhibit 9.8.

Coal mine

Normally this system will have structure SOS, since usually a stock of coal will exist at the pithead, output will be in anticipation of demand and all resources necessary for production will be held. Thus input stocks of consumed and non-consumed resources, and output product stocks will exist. Most of the input stocks will be non-consumed items such as labour, machinery, etc, although some consumed items such as fuel and indirect materials, e.g. props, etc, will be used.

Taking firstly the output, i.e. coal stocks, management will exercise control over input to these stocks, input or stock replenishment being on a near-continuous flow basis. Management will seek therefore to ensure that there is normally sufficient stock on hand to meet exceptionally high demands. In other words, a safety stock will be maintained to protect against periods when stocks begin to run down because demand rate is in excess of input rate. This will be an important factor in such a system because input rate, i.e. the rate at which the mine produces coal, cannot easily be altered at least not in the short term. However, because of seasonal factors demand rates may vary quite substantially. Another decision for operations management will therefore be the input or replenishment rate, i.e. the production rate of the mine. This rate will normally reflect estimated average demand from the customers to be served by the system.

The management of consumed item input stocks will resemble the situation described above. Again, management will exercise control over the input to stocks, such input flow being either intermittent or continuous. In the case of continuous flow, management will seek to determine the average or safety stock required to ensure that sufficient stock is on hand to cover periods of high usage in the mine, and additionally management will seek to ensure that the continuous replenishment rate of such stocks is consistent with the expected usage rate. The need for maintaining a safety stock arises from the inability to vary input flow rate sufficiently rapidly to match unexpected variations in usage rate. In some cases replenishment will be on an intermittent basis, in particular in cases where usage is low. Here management will seek either to establish a replenishment

SYSTEM	NORMAL STRUCTURE		INVENTORIES	TYPE OF ITEMS STOCKED	NATURE OF CONTROL OF THIS STOCK	TYPE OF FLOW CONTROLLED	INVENTORY MANAGEMENT DECISIONS	PRINCIPAL OBJECTIVES
Coal Mine	SOS	Input	Resources	Consumed	Input control	Intermittent or continuous	Determine stock replenishment level or interval; Determine average (or safety) stock, and stock replenishment rate	Customer service (as measured by the number or probability of stock outs)
				Non-consumed	Input control	Intermittent	Determine time of replacement	+
		Output	Products	Consumed	Input control	Usually Continuous	Determine average (or safety) stock, and stock replenishment rate	Resource productivity
				✕	✕	✕	✕	
Hospital Accident Ward	SCO	Input	Resources	Consumed	Input control	Intermittent or continuous	Determine stock replenishment level or interval; Determine average (or safety) stock, and stock replenishment rate	Customer service (i.e. to ensure no customer waiting or queueing time)
				Non-consumed	Input control	Intermittent	Determine timing of replacement	
		Output	(None)	✕	✕	✕	✕	
Taxi Service	SQO	Input	Resources	Consumed	Input control	Intermittent	Determine stock replenishment level or interval	Customer service (as measured by average customer waiting time or queue length)
				Non-consumed	Input control	Intermittent	Determine timing of replacement	+
		Output	Customers	Consumed	Output control	Intermittent (perhaps approaching continuous)	Determine stock levels or intervals at which output is to take place	Resource productivity (as for example measured by vehicle occupancy)
			(None)	✕	✕	✕	✕	

Exhibit 9.8 The nature of inventory management—examples

policy based on replenishments at a given stock level or replenishment at a given interval of time. The task therefore is to establish either the replenishment level taking account of possible variations in usage rate or replenishment, lead time, or to establish the interval for replenishment. In the case of the replenishment level system, management will seek also to identify the replenishment quantity which will normally be a fixed, i.e. batch, quantity whereas in the case of the fixed replenishment interval the replenishment order will often be sufficient to return stocks to a given maximum level.

The management of non-consumed resource stocks presents a somewhat different problem since here replenishment will be on a strictly intermittent basis, the objective being to replace items, e.g. pieces of equipment, labour, etc, when those items currently being used either fail, leave or complete their useful working life. The problem therefore is one of timing the replacement of such resources.

The principal objectives pursued in the management of inventories in situations described above will be firstly the provision of adequate customer service, and secondly the provision of adequate resource productivity. In this case, customer service will be measured by the number of stock-outs, or the probability of stock-outs, i.e. the probability of customer orders not being met from finished goods stocks. Because of the system structure which provides for the provision of output stocks, customer service from such systems will normally be high. Furthermore, resource utilization may also be high since the insulation of the function from demand fluctuations by means of output stocks permits a relatively level function rate and therefore a relatively high utilization. Inventory management policy will have a significant effect on the achievement of both objectives.

Hospital accident ward

Here the normal structure will provide for stocking of input resources without the need for customer queueing, i.e. structure SCO. Inventories will exist on the input side of the function only and will consist of consumed and non-consumed resources. There will be no stocks of finished goods nor, normally, stocks or queues of customers. Taking firstly the consumed item input stocks, management will seek to control such inventory through control of input flow, such flow being either intermittent or continuous. In the case of intermittent flow, management will seek to establish either a replenishment level or a replenishment interval and, in the case of the former, replenishment batch sizes. This is a classic inventory control problem involving intermittent input flow control. In the case of the control of continuous or almost continuous input flow, management will seek to establish a safety or average stock level sufficient to accommodate unexpectedly high demand, such safety stock being required since the input rate will be fixed in the short term and will reflect expected usage rate. Consumed item input stocks in hospital accident wards might consist of medications, bandages, disposable items, intermediate materials, direct materials, etc. In contrast non-consumed items will of course be all physical facilities, equipment, buildings, etc, together with labour, that is medical, nursing and administrative staffs. Here replacement is almost certainly likely to be on an intermittent basis, the task again being to determine the timing of replacement, such replacement being either replacement of resources lost or the renewal of exhausted resources. The objective in the management of such input resource inventories in the hospital accident ward situation will be principally concerned with the provision of adequate customer service. Here customer service will be considered satisfactory only insomuch as the system structure is maintained, i.e. only insomuch as customer queueing is avoided. A deterioration in customer service reflected in the queueing of customers will of course reflect a change in structure from SCO to SQO. In this latter case somewhat different inventory management problems will occur (as described below).

Taxi service

This system will normally operate with structure SCO, that is resources will be stocked for the provision of transport, and customers will not normally be required to wait for transportation.

Taking input resource stocks most of this stock will be of the non-consumable variety, e.g. taxi driver, etc, although some input resources will be consumed, e.g. fuel. Considering the management of the consumed input resources, control will be exercised on the input of such resources, input being by intermittent flow. Take for example, fuel, the manager of the system will input fuel intermittently as required or according to a predetermined plan. He will decide upon either a stock replenishment level or upon a stock replenishment interval. In either case this policy is likely to be somewhat flexible since opportunity will play a considerable part in this purchase of fuel. Nevertheless he will under normal circumstances seek either to purchase fuel when his available stocks fall to or below a certain level or alternatively he will perhaps purchase fuel (say) each morning before commencing his business. He might of course adopt a mixed policy, indeed in periods where demand is high he may require to replenish his fuel stocks when they fall towards zero, even though his normal policy is to replace them only on a daily basis. In the case of non-consumed input resources, replacement will again be on an intermittent flow basis and management decisions will be exercised upon input. Management will therefore decide upon replacement rather than replenishment, the objective being to replace items whose useful life has ceased or to replace items lost or damaged.

In some, perhaps unusual, cases the taxi will operate as structure SQO, i.e. customers will queue for transport. Such a situation may apply when a taxi has a monopoly on the supply of transport in an area and where demand is high. In the case of customer stocks or customer queues, management will exercise control over output. In such cases, the operations manager, that is the taxi driver, will have some control over output from the customer queue. Such output or depletion being on an intermittent basis. He will determine either the level at which the queue of customers is to be depleted, i.e. he may choose to pick up customers from a queue at a station or pick-up point when that queue has reached a certain length, or, as is more likely, he may chose to pick up customers at certain either fixed, or variable, intervals according to his availability.

His principal objectives throughout will be the provision of a satisfactory customer service and of course the provision of a relatively high resource productivity. Customer service in this case will be measured by average customer waiting time or average customer queue length, more likely the former. Resource productivity will be measured by the occupancy rate or the percentage of time the cab or cabs are occupied by customers. In this case inventory management will have relatively little effect on either criteria since capacity considerations and to some extent activity scheduling considerations will influence both customer service and resource productivity.

SUMMARY OF CHAPTER 9

The inventory management problem is related to that of activity scheduling as both deal with physical flows through the systems. Together they provide a chain of scheduling decisions between customers and suppliers.

Inventories can exist at *two stages* in the system—input and output. Input inventories can consist of two types of item—*consumable* or *non-consumable items*.

The replenishment and depletion of stocks can take place on a *continuous* or *intermittent flow* basis.

The complexity and likely effectiveness of inventory management depends upon the variability or *unpredictability of output* (*or input*) *levels* and upon the opportunity for and the *extent of control* of input (or output). Three basic types of consumable item inventory control situations can be identified, i.e.:

(1) Control of inputs
(2) No control
(3) Control of outputs

The nature of the inventory management problem will depend upon the strategy employed i.e.:

(1) The *location of control*, and
(2) The *nature of flow* (whether intermittent or continuous).

Section 4

The Nature and Procedures of Operations Management

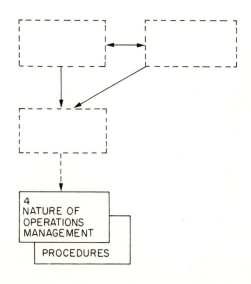

This section provides a brief introduction to some of the procedures and techniques available for problem solving in operations management. It is not our intention to describe them in detail, as this is adequately done elsewhere, but simply to consider conventional procedures in the three principal problem areas and to indicate where and when they may be used. (SOME READERS MAY WISH TO OMIT THIS SECTION)

CHAPTER 10

Procedures for Capacity Planning

In Chapter 7 it was argued that the planning and control of system capacity is perhaps the most important responsibility of operations management, i.e. the most important of the three principal problem areas.

Capacity planning has been shown to involve:

(1) The determination of the capacity required in the system.
(2) The development and implementation of the strategy for the use of given system resources to meet demand fluctuations.

The *control of capacity* has been shown to involve:

(3) The manipulation and deployment of given system resources. It has been argued that such control is achieved primarily through activity scheduling and control decisions and, indirectly, through inventory management.

The nature of capacity management is summarized in Exhibit 10.1. This chapter deals with procedures, methods, and techniques pertinent to capacity planning,

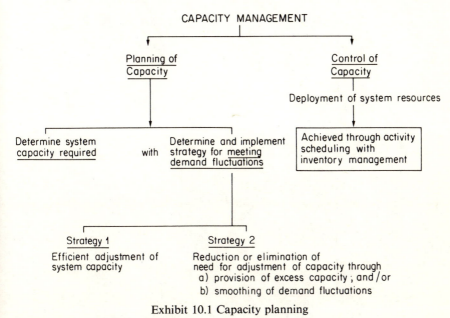

Exhibit 10.1 Capacity planning

firstly the determination of capacity requirements, and secondly meeting demand fluctuations, whilst Chapters 11 and 12 deal, in effect, with the control of system capacity. We provide no more than a brief survey or guide. Nothing is discussed in detail. Our objective is simply to introduce procedures, which in general are adequately discussed elsewhere, and to indicate their relevance and limitations for decision making in operations management. This is intended as a link between the concepts discussed previously and problem solving and decision making procedures examined in most operations management texts. References are given at the end of the chapter to facilitate access and further study.

CAPACITY REQUIRED—1: SOME CONSIDERATIONS

The objective in this aspect of capacity planning will be the determination of required capacity through either the measurement or estimation of the demand to be placed on the system. Estimation or forecasting of future demand will normally be necessary where resources and/or output is stocked, i.e. in structures SOS, DOS, SOD, SCO and SQO. In other situations, i.e. structures DOD and DQO, since no output or resources are provided prior to receipt of customer order demand can be measured, estimation is therefore unnecessary and the capacity planning problem is considerably simplified.

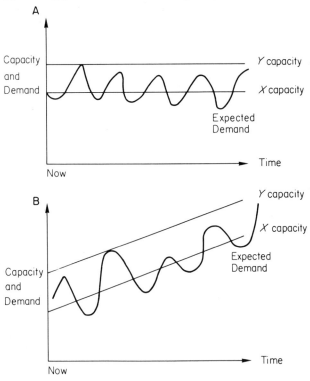

Exhibit 10.2 Demand forecasts of planned capacity

When forecasting future demand for capacity planning purposes fluctuations will be expected but to some extent ignored. For example, where there is no trend in expected demand, a capacity equal to average expected demand might be provided, i.e. capacity X in Exhibit 10.2A. If, however, a strategy of providing excess capacity is to be adopted a higher capacity may be provided, e.g. capacity Y in Exhibit 10.2A. Where demand is expected to increase or fall, e.g. Exhibit 10.2B, correspondingly increasing or reducing capacity may be provided, again taking account of the strategy adopted for the use of system resources, e.g. capacities X or Y in Exhibit 10.2B. Determination of system capacity requires consideration of the strategy for meeting demand fluctuations.

Demand forecasting[1] (structures SOS, DOS, SOD, SCO, SQO)

The length of the forecast period will depend largely upon the nature of system resources and the nature of the market. Capacity plans may involve periods in excess of five years where there is sufficient stability or predictability of the nature of demand (i.e. *what* the customer requires). A long-term view may be essential where there is a long lead time on the provision or replacement of resources. Examples might include:

Manufacture: Steel manufacture
Electricity generation and supply
Oil production
Transport: Airlines
Rail systems
Service: Hospitals
Telephone service

In contrast a shorter term view would be appropriate where the nature of demand is less stable or less predictable, and where resources are more readily provided or replaced, or where the manner in which the function is accomplished may change, through for example technological change. Examples might include:

Manufacture: of fashion goods; consumer durables
Supply: retail shops; mail order
Transport: bus service; taxi service; road delivery service
Service: secretarial services; security service

In certain cases, namely where the *nature* of future demand is unknown—corresponding to structures DOD and DQO—future demand cannot be estimated. Here capacity planning is simpler and the future time period zero. In fact capacity planning as described above and in Chapter 7 does not occur since demand can be *measured* and appropriate capacity provided. However in most cases there will remain some uncertainty in determining required capacity since every customer request will be different. It will therefore be necessary to provide for some capacity level adjustment.

In forecasting demand for goods or services it is appropriate to recognize that in many cases demand is a function of time. The classic life cycle curve[2] shown in Exhibit 10.3 illustrates one such relationship. This curve will normally apply in the case of goods or services consumed directly by the public (e.g. domestic appliances, sport and leisure facilities, etc.). The time scale will depend upon

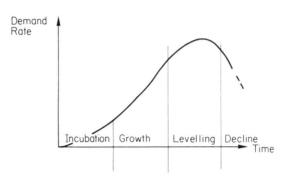

Exhibit 10.3 A life-cycle curve

the nature of the product or service. Whilst there is a tendency to assume continued growth for goods and services such as raw materials, basic services, transport, etc (e.g. steel, fuel, medical care, rail and air transport), a similar life cycle relationship may in fact exist although the time scale may be considerable. Demand for energy, particularly electricity, has recently begun to exhibit the characteristics of an S-curve, after a considerable period of apparently exponential growth. The use of canal and rail transports show this relationship. Various mathematical formulae exist for the description of such curves. Exponential functions are often employed to describe the incubation and growth periods. Various policies might be employed for the provision of capacity to satisfy such demand. At one extreme, sufficient capacity to meet all expected demand might be provided from the outset, with attendant benefits of econo-

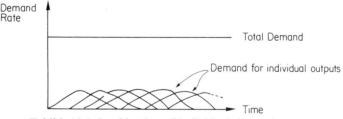

Exhibit 10.4 Combination of individual demand curves

mies of scale in ordering, acquisition, training, etc. Alternatively, capacity might be matched to demand by incremental change over time, with benefits in utilization, etc. Systems providing several outputs from common resources might, despite the life cycle characteristics, expect relatively steady total demand. Exhibit 10.4 illustrates.

Forecasting techniques are adequately covered elsewhere. Their application in forecasting demand for capacity planning is relatively straightforward and since there is no necessity to provide accurate predictions of short-term fluctuations, the emphasis will be upon demand levels and trends.

Aggregation

The term *aggregate planning*[3] is often employed in the capacity context. The implication is that such planning is concerned with total demand, i.e. all demands collected together. This is of relevance only in multi-channel systems where different goods or services are provided, and in such cases aggregate or capacity planning will seek to estimate or measure all demands and express the total in such a way as to enable sufficient of all resources (or total capacity) to be provided. Demand for all outputs must therefore be expressed in common capacity related units such as the number of resources or resource hours required. An operating system may for example provide three types of service, or service three types of customer. The estimated demand for each source expressed in, for example, hours per unit time (e.g. week) required for each type of resource must be identified and totalled. Given this aggregate demand a sufficient quantity of each resource can be provided.

Resource improvement and deterioration

In determining the capacity, i.e. the quantity of resources required to meet either forecast or measured demand, it should be noted that the capability, or capacity, of a given set of resources might also change with time. The reliability[4] of machinery might change (Exhibit 10.5 shows a typical reliability curve) and

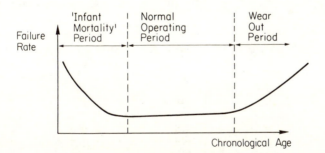

Exhibit 10.5 A reliability curve

the efficiency of labour might improve, due to the learning effect.[5] It is sufficient for our purposes to consider learning to be the process by means of which an individual or an organization acquires skill and proficiency at a task which in turn permits a higher task performance (e.g. shorter time required for completion). Exhibit 10.6 shows a typical learning or improvement curve. Clearly

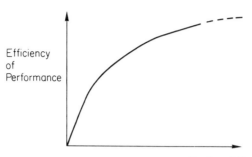

Number of Repetitions of Task

Exhibit 10.6 Learning curve

the effect of such learning is to increase the capacity of a given quantity of labour resource. Such capacity change effects may be of considerable importance in capacity planning.

CAPACITY REQUIRED—2: CUMULATION

The capacity provided to satisfy estimated or measured demand will, as shown above (Exhibit 10.1), be influenced by the strategy employed for meeting demand fluctuations. The use of *cumulative curves*[6] is a method for examining alternatives.

Exhibit 10.7 gives the estimated monthly demand for a one year period. The figures are plotted cumulatively in Exhibit 10.8 which also shows two possible cumulative capacity curves. Curve 1 corresponds to a capacity of 37·5 resource hours per day—the minimum required to ensure that capacity is always equal to or in excess of expected demand for this period. The adoption of a strategy of

Month	Working days	Cumulative days	Estimated demand (in resource hours)	Cumulative estimated demand
Jan	20	20	500	500
Feb	18	38	650	1150
Mar	22	60	750	1900
Apr	18	78	900	2800
May	21	99	700	3500
June	20	119	500	4000
July	20	139	300	4300
Aug	10	149	300	4600
Sept	20	169	450	5050
Oct	21	190	500	5550
Nov	20	210	550	6100
Dec	18	228	300	6300

Exhibit 10.7 Estimated monthly demand

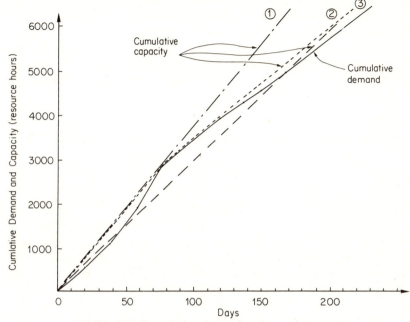

Exhibit 10.8 Cumulative demand and capacity curves

providing sufficient excess capacity to eliminate the need for capacity adjustment (strategy 2a, Exhibit 10.1) would lead to the provision of such capacity. The provision of approximately 30 resource hours per day—curve 2—might result from the adoption of a different strategy for the use of resources (strategy 2b, Exhibit 10.1). Such an arrangement would give rise to either increasing output stock (structures SOS and DOS) or reducing customer waiting time (structures SOD, DOD, DQO and SQO), during the period up to day 50 and after day 160, when capacity exceeds demand. Day 50 to day 160 would see either:

(1) Stock diminishing or depleted (structures SOS and DOS)
and/or
(2) Increased customer waiting time (structures SOD, DOD and DQO)
and/or
(3) loss of trade (structures SOD, DOD, SCO, DQO and SQO)

since capacity would be lower than expected demand.

Both curves 1 and 2 require no capacity adjustment during this period. The adoption of an approach relying wholly upon or involving the strategy of providing for efficient adjustment in capacity (strategy 1, Exhibit 10.1) might give rise to the provision of capacity in the manner of curve 3 in which one capacity adjustment is made (at day 75) and which provides for the satisfaction of all forecast demand without the use of output stocks, customer queueing or loss of trade, yet with better capacity utilization than curve 1.

In practice the use of cumulative graphs in capacity planning must take account of the lead time normally required between the use of capacity or resources and the satisfaction of customers. For example, the expected customer demand in April equivalent to 900 resource hours (Exhibit 10.7) would necessitate the provision of appropriate capacity at an earlier period. Hence the capacity curves in Exhibit 10.8 should in fact be displaced forward by the amount of this lead time.

MEETING DEMAND FLUCTUATIONS—1: STRATEGY ASPECTS

The strategy employed for the use of resources will involve either:

(1) the provision of efficient capacity adjustment, or
(2) the elimination or reduction of the need for capacity adjustment, or both.

The feasibility of each strategy will largely reflect the structure of the system (see Exhibit 7.4 and 7.5). In this section we shall examine the means through which these strategies might be implemented and the factors which will influence their relative merit and cost. Finally we shall introduce and comment on the relevance and application of some quantitative methods.

Adjustment of capacity

Some aspects and implications of the strategy of adjusting system capacity are considered below.

Make or buy/subcontract

An organization may find another willing to take some of the excess demand. This will often involve a higher cost per unit of output because of subcontractor overheads and profit and increased cost of inspection, administration, transport, etc. Such an approach is least reliable, most expensive and least flexible when it is needed most, since a need for greater capacity often is associated with a general increase in total 'industry' demand. At such time, potential subcontractors will also be busy.

Workforce and hours changes

All system outputs will not necessarily be subject to high demand simultaneously. If labour can move from one function/system to another and machinery is also flexible, high demand for one output may be offset by low demand elsewhere. Again this approach may be least reliable when most needed for the reasons given above. The additional costs incurred derive from the time and effort invested in multijob training and lower performance during learning periods.

Changes in labour force and working hours are the normal means for adjusting system capacity. The most used method for increasing labour capacity is overtime working. Overtime and shift premiums are added costs and productivity may be lower, supervision and service costs higher, etc. Long hours may also lead to more accidents, illness and greater absenteeism. Where layoffs are undesirable, overtime working is preferable to adding temporary workers. Adding workers to the payroll increases the costs of recruitment and training. Employment often can be reduced without formal layoffs. Normal labour turnover may help reduce labour capacity. Again this approach may be unavailable when most required. An alternative to layoffs is shorter work weeks or idle time. The latter requires the company to carry the cost of underutilized capacity.

Deferred maintenance

In periods of temporary high demand, it is possible to keep resources operating longer by not closing down as scheduled. Demand reductions will permit shutdowns earlier than originally scheduled. The costs involved in such an approach may derive from the difficulties of scheduling the use of maintenance facilities, earlier breakdowns, etc.

Activity scheduling

The selection of the appropriate activity schedules, including if possible the scheduling of customer arrivals can contribute considerably to the ability of the system to meet demand fluctuations. Such changes which will not substantially affect capacity, are relevant only for capacity increase. Costs incurred derive from the increased complexity of scheduling and control, perhaps to some extent offset by higher capacity utilization.

Avoidance or reduction of need for capacity adjustment

Some aspects and implications of this strategy are examined below.

Refusing business, reducing service and adjusting backlogs

An organization may decline an order when its capacity is fully utilized. If capacity is short throughout an industry the organization may be able to lengthen its deliveries, and therefore its customer queues without loss of trade. In such cases the customer's order may simply be added to the backlog of work. In the opposite situation reduction of the backlog is desirable. Given customer queueing during periods of high demand, greater flexibility may exist in selecting orders to fill in the gaps in activity schedules. This may give better capacity utilization but poorer customer service, which may be tolerated if competitors offer similar delivery at comparable price.

Adjustments in inventory levels

An output stock permits the utilization of capacity during low demand periods and relieves the congestion, helps avoid queueing, etc, during peak loads. In some cases, only partly completed output may be stocked, and finished on receipt of order. Customer service might be improved and this may bring an increased share of the demand, greater workloads and further capacity problems. The costs of inventory are related primarily to the opportunity costs of the capital invested and the risks of obsolescence.

Changing price levels

To maintain operations, prices can be manipulated especially on goods and services where individual price quotations are offered subject to negotiation. When demand is high prices can be increased to increase total contribution, and in total low demand periods, prices can be lowered toward variable costs to help reduce the fall in demand placed on the system.

MEETING DEMAND FLUCTUATIONS—2: QUANTITATIVE METHODS

Many quantitative methods are of some value in capacity planning. Here we need only consider those of relevance to decision making in developing and implementing strategies for meeting demand fluctuations. Methods concerned with decisions in specific areas only, e.g. inventory management and scheduling techniques, will be examined in Chapters 11 and 12. We shall review those methods which deal with several aspects of one or both basic strategies (Exhibit 10.9). For example certain methods seek to determine the best (least cost) combination of output rate, inventory level, etc, whilst others are concerned with the minimization of capacity, capacity change, inventory and lost order costs. The remainder of this chapter therefore provides a brief, annotated review of the best known methods, the majority of which were developed for use in manufacturing systems.

The reaction rate[6]

$$\left\{ \begin{array}{l} \text{Strategy 1: Capacity adjustment} \\ + \\ \text{Strategy 2: Use of Inventories bII} \end{array} \right\} \quad \left\{ \begin{array}{l} \text{Structures} \\ \text{SOS and DOS} \end{array} \right\}$$

The reaction rate of adjustment of capacity determines how much capacity adjustment is made to meet a forecast fluctuation in demand. Capacity adjustment by the full amount of demand variation gives a 100 per cent reaction rate, hence fluctuations in demand are transmitted directly to function. This might lead to severe problems and costs in acquiring or disposing of resources, e.g. personnel. Reaction rates can take any value from zero to 100 per cent. Low reaction rates lead to stable operation but higher inventories, and higher reaction rates would lead to the reverse conditions. Short review periods between decisions lead to small capacity fluctuations and smaller inventories for a given reaction rate.

The selection of a reaction rate and review period depends upon the balance of inventory carrying costs in relation to the costs of capacity change. The mathematical analysis of the reaction rate approach indicates the optimum combination for a given application. If much of the fluctuation in demand derives from random causes, there is little need to react quickly. Lower reaction rates and short review periods will cause small adjustments fairly often. In

STRATEGY	1	2			
			a	*b*	
			Provide excess capacity.	I Fix upper capacity limit with effect of:	II Use stock to absorb demand level fluctuations
				loss of trade / customer queueing/waiting.	
Objective	Provide for efficient adjustment to demand level changes.	Eliminate or reduce need for adjustment in system capacity.			
Aspects	Make or buy decision/subcontract. Workface size. Hours worked. Activity Schedules. Maintenance Schedules.			Refuse business. Price changes. / Service level (backlogs)	Inventory levels

Exhibit 10.9 Meeting demand fluctuations—strategies and variables

general, high reaction rates and long review periods tend to overcontrol. The optimal amount of control minimizes the sum of the two types of costs, rather than minimizing one at the expense of the other.

Capacity smoothing—1: The transportation (LP) method[7]

$$\left.\begin{cases} \text{Strategy 1:} \\ \qquad + \\ \text{Strategy 2: bII Inventories} \end{cases}\right\} \quad \begin{cases} \text{Structures} \\ \text{SOS and DOS} \end{cases}$$

The transportation method of linear programming provides a means for minimizing a combination of capacity and inventory costs. Various alternative means for providing capacity are recognized, typically normal working, overtime working and subcontracting. The use of the method requires that demands for each of several periods are satisfied from inventory and/or from the use of normal plus if necessary additional capacity, in such a manner that total cost is minimized.

Demand during period 1 can only be satisfied from stock. Period 2 can be satisfied from stock, regular or overtime capacity, etc, through to final stock which will consist of all opening stock plus output which has not been used to satisfy demand in the four periods.

This approach suffers one major disadvantage in being static, i.e. it is a method for formulating policy for a given period assuming no changes in circumstances, etc, during that period. Repeated recalculations for a forward period will overcome this to some extent but there is a danger of suboptimization.

N.B. Several other LP formulations have been employed.[8] They offer a similar capacity smoothing approach to that considered above, but might additionally take account of loss of trade, the cost of changing capacity, etc (systems SOS, DOS, SOD, DOD, SCO, DQO, SQO).

Capacity smoothing—2: The linear decision rule[9]

$$\left.\begin{cases} \text{Strategy 1:} \\ \qquad + \\ \text{Strategy 2: bII Inventories} \end{cases}\right\} \quad \begin{cases} \text{Systems} \\ \text{SOS and DOS} \end{cases}$$

The objective of this sophisticated approach is the derivation of a series of linear equations (or 'decision rules') for use in specifying the output rate and capacity required to meet forecast demand and minimize total costs. The costs of carrying inventory, payroll costs (including overtime) and the cost of changing the size of the labour force are considered.

'Rules of thumb'—heuristic methods[10]

$$\left.\begin{cases} \text{Strategy 1:} \\ \qquad + \\ \text{Strategy 2: bII Inventories} \end{cases}\right\} \quad \begin{cases} \text{Systems} \\ \text{SOS and DOS} \end{cases}$$

A 'heuristic' method provides a good, but not necessarily the best, solution. In reality most operations management decisions are heuristic or, more colloquially, 'rules of thumb'.

The management coefficients model uses a simplified version of the workforce production level decision rules incorporated in the linear decision rule method. Coefficients for these rules are determined by regression analysis on historical performance, i.e. managers' actual past behaviour. The equation is then used to indicate the future decision as with the linear decision rules.

'Search' methods[11]

$$\left\{\begin{matrix} \text{Strategy 1:} \\ + \\ \text{Strategy 2: bI, II} \end{matrix}\right\} \quad \left\{\begin{matrix} \text{Structures} \\ \text{SOS and DOS} \end{matrix}\right\}$$

An alternative to the heuristic approach involves a form of search of feasible solutions to establish the coefficients for the linear decision rule method. The coefficients are established to minimize total cost and the two equations are then used for planning purposes. Two approaches have been developed. The parametric production planning model deals with inventory, work force changes, overtime costs and customer queueing. The search decision rule is of a similar nature.

Simulation methods[12]

It has been pointed out that heuristic methods result in fairly simple decision rules for planning. In contrast simulation models are used when such simple procedures are not appropriate. Heuristic methods are perhaps more appropriate for single channel systems elsewhere, and particularly in the case of push systems the simulation approaches might be more appropriate.

SUMMARY OF CHAPTER 10

Capacity planning, the principal aspect of capacity management, involves:

(1) the determination of the capacity required in the system, and
(2) the development and implementation of the strategy for the use of given system resources to meet demand fluctuations.

Determination of required capacity necessitates:

(1) *Demand forecasting* for structures SOS, DOS, SOD, SCO and SQO, and the *measurement* of demand for structures DOD and DQO.
(2) The *aggregation* of demands for multi-output systems.
(3) Consideration of factors giving rise to resource *improvement and deterioration* e.g. the effects of learning and reliability.

The use of *cumulative demand* and capacity curves provides a useful means for examining alternative strategies under (1) and (2) above.

Capacity adjustment to meet demand changes might be achieved through decisions in respect of:

(1) Make or buy and subcontracting
(2) Workforce and hours
(3) Maintenance
(4) Scheduling

Avoidance of capacity adjustment might be achieved through:

(1) Refusing business, reducing service or adjusting backlogs
(2) Adjusting inventory levels
(3) Changing price

The various *quantitative methods* which are available for capacity planning through one or both of these strategies are of relevance in a limited range of situations, e.g.:

	Strategy	*Structure*
The reaction rate method	1 and 2	SOS, DOS
The transportation method	1 and 2	SOS, DOS
The linear decision rule	1 and 2	SOS, DOS
Heuristic methods	1 and 2	SOS, DOS
Search methods	1 and 2	SOS, DOS

REFERENCES

1. Brown R. G., *Statistical Forecasting for Inventory Control*, McGraw-Hill, 1959.
2. Kotler, P., *Marketing Management*, Prentice Hall, 1967.
3. Buffa, E. S., *Modern Production Management*, 4th Ed., Wiley, 1973, Chapter 17.
4. Lewis, R., *An Introduction to Reliability Engineering*, McGraw-Hill, 1970.
5. Holding, D. H., *Principles of Training*, Pergamon, 1965.
6. Buffa, E. S., *Modern Production Management*, 4th Ed., Wiley, 1973, Chapter 17.
7. Bowman, E. H., 'Production Planning by the Transportation Method of Linear Programming', *Journal of Operations Research Society* (Feb. 1954).
8. Hanssmann, F. and Hess, S. W., 'A Linear Programming Approach to Production and Employment Scheduling', *Management Technology*, **46** (January 1960).
9. Holt, C. C., Modigliani, F., Muth, J. and Simon, H., *Planning Production Inventories and Work Force*, Prentice Hall, 1960.
10. Bowman, E. H., 'Consistency and Optionality in Managerial Decision Making', *Management Science*, **9**, No. 2 (1963).
11. Taubert, W. H., 'A Search Decision Rule for the Aggregate Scheduling Problem', *Management Science*, **14**, No. 6 (1968).
12. Vergin, R. C., 'Production Planning under Seasonal Demand', *Journal of Industrial Engineering* (May 1966).

CHAPTER 11

Procedures for Activity Scheduling

The external and internal orientations of operations scheduling have been identified in Chapter 8. Externally-oriented scheduling is primarily concerned with the customer service objective whilst internally-oriented scheduling is more likely to focus upon resource productivity. Externally-oriented operations scheduling decisions must take direct account of customer demands, and in many cases such external requirements will be accommodated through inventory decisions. In all but structure DOD systems (function from source direct to customer—see Exhibit 8.2), some scheduling decisions will be largely internally oriented.

We have argued (Chapter 7) that scheduling decisions are one means for the management of capacity, in particular the control of the system capacity. We have seen in Chapter 8 that the nature of the scheduling problem will depend to some extent on the type of system structure which exists. In this chapter we will again refer to such issues. We shall seek to identify the procedures which are of value in *activity scheduling* and will examine their relevance and limitations with particular reference to external/internal orientation, capacity management, the nature of the scheduling problem and system structure.

The procedures conventionally employed in activity scheduling are listed in Exhibit 11.1. They are discussed in general terms below and in relation to each of the above issues in subsequent sections.

PROCEDURES FOR ACTIVITY SCHEDULING

Reverse scheduling

External due date considerations will influence activity scheduling in certain structures (SOD, DOD and SCO). The approach adopted in scheduling activities in such cases will often involve a form of reverse scheduling of the Gantt chart type illustrated in Exhibit 8.4. The procedure provides for a visual display of activity durations, but suffers the limitation of not showing sequential interdependence of activities. A form of a reverse scheduling from a given or required due date is however a conventional procedure for activity scheduling using both manual and computer techniques.

| | ORIENTATION | | SYSTEMS STRUCTURE |
	EXTERNAL	INTERNAL	
Reverse scheduling	√	√	SOD, DOD, SCO
Gantt charts	√	√	SOD, DOD, SCO
Network scheduling	√	√	SOD, DOD, SCO
Resource aggregation	√	√	SOD, DOD, SCO
Resource allocation	√	√	SOD, DOD, SCO
Dispatching/sequencing	√	√	SOS, SOD, DQO, SQO
Assignment		√	SOS, SOD, DQO, SQO
Batch scheduling		√	SOS, DOS, DQO, SQO
Timetabling	√	√	SOS, DOS, SCO, DQO, SQO

Exhibit 11.1 Procedures for activity scheduling

Network scheduling[1]

In more complex situations network scheduling might be appropriate since it provides a means for establishing schedules for sequentially interdependent activities taking account of precedence constraints and in some cases resource constraints. In its simplest form the application of network analysis might involve forward scheduling for the purpose of identifying the minimum normal duration for a project and the activities critical to that duration. Alternatively, given a required due date (i.e. the required duration), the procedure can be used to identify required start dates. In either case the quantity of resources required for each period might be calculated or *aggregated*. The latter approach may indicate that a due date is infeasible. In such cases extra capacity (i.e. resources) may be needed. Alternatively, given fixed resources, activities might be scheduled in order to smooth the requirements for each resource in each time period. Such an approach is often referred to as resource *allocation* or smoothing. Exhibit 11.2 indicates the type of procedure involved.

Dispatching[2] and sequencing[3]

Both procedures are of relevance in determining the priority order for available jobs or customers through given resources or facilities. Their use therefore is of relevance in the scheduling of activities in all systems where input stocks exist, excepting structure SCO. Sequencing procedures seek to establish an order for all available customers or jobs, whilst the dispatching approach provides a solution via the use of local rules, e.g. 'first come first served', 'shortest

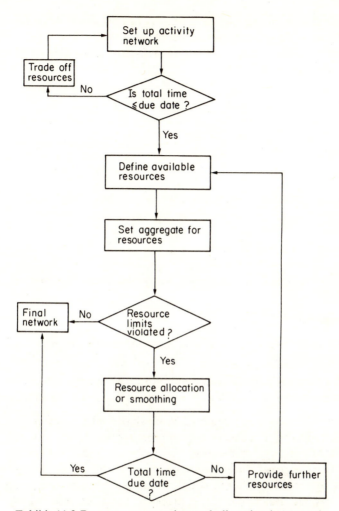

Exhibit 11.2 Resource aggregation and allocation in network
scheduling

processing time', 'earliest required due date', etc, for the selection from the
queue or stock available. 'First come first served' dispatching rules are, for
example, adopted in the scheduling of customers in hospital accident wards,
G.P. consultations, and transport systems, etc. 'Earliest due date' criteria
might be used in maternity wards!

Dispatching rules might be chosen in order to maximize due date delivery
performance, tardiness of jobs, waiting time, etc.

Assignment[4]

Various approaches exist, often depending upon some linear programming
formulation, for the solution of scheduling problems involving the allocation

or assignment of available jobs or customers against available facilities. Exhibit 11.3 indicates a job or customer schedule for a five-by-five type problem, i.e. five customers against five facilities, each facility being able to process each customer at a given cost.

```
                        Resource
                A   B   C   D   E
            1 ┌ 5   6   4   8  ③ ┐
            2 │ 6  ④   9   8   5 ◄──── Cost associated with the
  Customer  3 │ 4   3  ②   5   4 ◄     assignment of given customer
            4 │ 7   2   4  ⑤   3 │     to given resource
            5 └③   6   4   5   5 ┘
```

N.B. Least cost assignment
      = 1:E, 2:B, 3:C, 4:D, 5:A

Exhibit 11.3 Assignment matrix (least total cost solution)

Batch scheduling[5]

Basically three types of batch scheduling problem exist, namely:

(1) Batch processed once only,
(2) Batch processed repeatedly at regular intervals when needed,
(3) Batch processed repeatedly at known intervals to satisfy continual demand.

The problem of establishing required batch sizes applies in cases of (2) and (3), that of determining a sequence for batches applies in the case of (3), whilst batch scheduling programmes occur in cases (1) and (2). Classic batch size formulae (see Chapter 12) are conventionally employed in calculating batch sizes, batch size again being influenced by cost of acquiring or setting up facilities, storage costs, etc. Batch sequencing problems are often solved using the linear programme assignment techniques in a similar manner to that illustrated above, whilst the scheduling of batches is analogous to the type of scheduling already referred to and might involve the use of Gantt charts, reverse scheduling, network analysis, etc.

Timetabling

The timetabling of activities is of particular relevance in respect of repetitive functions. Bus, train, and air services usually operate to a timetable. Similarly the activities of certain service systems are timetabled, e.g. cinemas. These are customer push systems, hence customers are required to take advantage of the function at predetermined times. The function is not performed at other times, hence customers arriving at the wrong time must wait, and of course, an absence of customers at the time selected for the performance of the function, or the availability of insufficient customers to fully utilize the facilities provided, will necessarily give rise to underutilization of capacity. In many situations time-tables are necessary, since common resources are deployed to provide a variety or a series of functions. In many transport systems, for example, vehicles travel a set route providing movement for individuals between points along that route. In certain service systems, for example, hospitals, common resources such as specialists provide a service in a variety of departments, or for a variety of types of customer in a given period of time. Certain out-patient clinics operate in this fashion. In all such cases a timetable will normally be developed and made available to customers. Much the same situations may apply in customer pull systems where function occurs at given times. It follows that in all such cases the nature of demand must be predictable, since in effect function is undertaken in anticipation of demand. In fact, in such cases, the absence of (sufficient) customers at the time selected for the performance of the function may give rise to the creation of output stock. Alternatively, output may be lost or wasted. The timetabling of such activities is an exercise in internally-oriented scheduling, since no direct account is taken of individual customer's demands. The development of timetables will take account of the time required for, or the duration of, activities, and in many cases (e.g. transport systems) the required or preferred order or sequence.

Few quantitative techniques are relevant in the development of such time-tables. 'Routing', flow planning, and vehicle scheduling procedures are of relevance in timetabling transport systems.[6] In some cases the problem will resemble that of sequencing outlined above, whilst in some situations it may be convenient to utilize Gantt charts, and simulation procedures in both the development and evaluation of timetables.

CRITERIA AND EMPHASIS

In most situations it will not usually be feasible, or in some cases desirable, to attempt to develop optimum activity schedules. Because of the dynamics of situations and the unpredictability of demand, a degree of control is essential. The relative importance attached to scheduling or control will depend to some extent on the type of situation and particularly the type of structure existing. In general, scheduling will be more complex in function to order situations where scheduling decisions will be required directly to absorb external (i.e. demand) fluctuations. Furthermore, in such situations the degree of function repetition may be less, therefore the need for control may be greater. In contrast, in function to stock situations, scheduling will be somewhat easier, therefore the need for control somewhat less. In demand push situations where stocks exist between function and customer, scheduling will tend to be easier and the need for control less. Demand levels may also influence the relative complexity and importance of scheduling and control activities. In general, high demand level will be associated with function repetition and the provision of special purpose resources together with product or service specialization. In contrast relatively low demand levels may be associated with relatively low function repetition, high product or service variety and the use of general purpose resources, hence scheduling may be complex and there will be a relatively greater dependence upon control. In situations where demand predictability, i.e. the nature of demand, permits the provision, i.e. stocking, of resources and where demand levels are high, the provision of special purpose equipment may give rise to an emphasis upon the provision of balance, the avoidance of interference and consideration of learning or improvement effects. In such situations much of the internally-oriented scheduling will be 'built in' to the system. In contrast, in situations where demand predictability is low and where demand levels are low also, accurate scheduling will be impossible, hence low equipment utilization, high work in progress, and/or customer queueing will be evident. In these situations the use of local dispatching rules, resource smoothing, allocation of jobs, etc, together with an emphasis upon control will be evident.

Criteria or measures of effectiveness of activity scheduling and control might include the following:

(1) The level of finished goods or work in progress (structures SOS and DOS).
(2) Percentage resource utilization (all structures).
(3) Percentage of orders delivered on or before due date (structures SOD, DOD and SCO).
(4) Percentage stock-outs/shortages (structures SOS and DOS).
(5) Number of customers lost (all structures).
(6) Down time/set ups, etc (all structures).
(7) Customer queueing times (structures SOD, DOD, DQO and SQO).

In virtually all cases there will be a need to avoid suboptimization, that is it will be easy in most cases to satisfy each of the above criteria individually. However, as we have pointed out in Chapter 3, the objective of operations management, and therefore an objective of activity scheduling, would be to obtain a satisfactory balance between customer service criteria and resource utilization criteria.

SCHEDULING VS. CAPACITY AND STRUCTURE

External or due date scheduling considerations will predominate in structures SOD, DOD and SCO. In such situations reverse or Gantt chart scheduling might be employed although in more complex situations, network techniques might be appropriate. In both cases scheduling decisions may affect and be affected by capacity factors. Thus activities may be scheduled to permit completion within given resource constraints; alternatively, in order to achieve external, i.e. due date, requirements, capacity may be adjusted. The use of queueing techniques in respect of externally-oriented scheduling is of relevance largely where controllable customer input situations exist.

Internally-oriented scheduling will predominate in structures SOS, DOS, DQO and SQO. In such situations dispatching, sequencing, batch scheduling, timetabling and assignment procedures might be employed. In such structures a capacity management strategy requiring minimum capacity adjustment will normally apply, hence activity scheduling and capacity management considerations will not be closely interrelated.

SUMMARY OF CHAPTER 11

Various *procedures* are available for scheduling activities to satisfy external or internal requirements.

In general scheduling will be more complex in *function to order* situations, where additionally there will normally be a greater need for control. Such situations will normally be characterized by low repetition and low demand levels.

Criteria for measuring the effectiveness of scheduling might include:

(1) Level of finished stock or work in progress (structures SOS and DOS).
(2) Percentage resource utilization (all structures).
(3) Percentage of orders delivered on or before due date (structures SOD, DOD and SCO).
(4) Percentage stock-outs/shortages (structures SOS and DOS).
(5) Number of customers lost (all structures).
(6) Down time/set ups, etc (all structures).
(7) Customer queueing times (structures SOD, DOD, DQO and SQO).

REFERENCES

1. Wild, R., *Techniques of Production Management*, Holt, Rinehart and Winston, 1971, Chapter 9.
2. Buffa, E. S., *Modern Production Management*, (4th Ed) Wiley, 1973, Chapter 19.
3. Eilon, S., *Elements of Production Planning and Control*, MacMillan, 1962, Chapter 13.
4. Wild, R., *Techniques of Production Management*, Holt, Rinehart and Winston, 1971, (Appendix 1).
5. Wild, R., *Mass Production Management*, Wiley, 1972, Chapter 7.
6. Mitchell, G. H., *Operational Research*, English University Press, 1972, Chapter 5.

CHAPTER 12
Procedures for Inventory Management

The nature of inventory management was examined in Chapter 8, where it was noted that one important function of inventories is the accommodation of short-term differences between input and output sales. It is important to note that such differences might result from fluctuations in output rate or input rate or both. A distinction was made between output and input control situations. The former often occurs in input push systems where customers queue and where there may be little direct control of input, but with the possibility of exercising some control of output. A further distinction was made between intermittent and continuous flow situations, the former requiring decisions on the timing and amount of stock replenishment, and the other on rates of replenishment. Quantitative techniques tend to abound in conventional treatments of inventory management. It is appropriate, therefore, that in this review of procedures we attempt to place such techniques into the context developed in Chapter 8 and in doing so further describe the types of inventory management problem likely to be encountered in operations management.

COSTS IN INVENTORY MANAGEMENT

Two sets of parameters are of immediate relevance, namely inventory costs and customer service costs.

Inventory costs

Items held in stock incur costs. *Holding costs* comprise the costs of storage facilities; provision of services to such facilities, insurance on stocks, costs of loss, breakage, deterioration, obsolescence and the opportunity cost or notional loss of interest on the capital tied up. In general, an increase in the quantity of stocks held will be reflected in an increase in holding costs although the relationships may not be linear. For example, costs for increasing stock holdings may be in the form of a step function, since increased space is required when stocks reach certain levels. The cost of capital, insurance, etc, may also be discontinuous through the effect of price breaks or quantity discounts. Stock holdings of a certain level may permit replenishment in quantities sufficient to attract quantity discounts. Other things being equal, higher costs of holding will result

134

		INVENTORY COST		CUSTOMER SERVICE COSTS	
		Holding costs	Inventory change cost	'Starving' cost	'Blocking' cost
Intermittent flow	Input control	Cost of space, insurance, tied up capital, loss and deterioration etc.	Cost of a replenishment	Cost of stock-outs (i.e. inventory empty)	N/A
	Output control		Cost of a depletion	N/A	Cost of customer queueing (i.e. inventory full)
Continuous flow	Input control		Cost of change of input flow rate	Cost of stock-outs (i.e. inventory empty)	N/A
	Output control		Cost of change of output flow rate	N/A	Cost of customer queueing (i.e. inventory full)

Exhibit 12.1 Cost parameters in inventory management

in lower stock quantities and vice versa. Certain *stock change costs* apply particularly in intermittent flow systems, e.g. in input control systems change costs will consist of the cost of ordering replenishment, and in some cases the cost of delivery of replenishment, items and the cost of receipt, inspection, placing in stock, etc. In output control systems change costs will constitute cost of ordering or initiating depletion and the cost of despatch etc.

Customer service

Customer service considerations influence inventory decisions in both input and output control systems. In input control systems, i.e. normally customer pull systems, service might be measured in terms of the number of occasions over a given period on which customer orders cannot immediately be satisfied from stock, i.e. the number of 'stock-out' situations. Equally the probability of such stock-outs might also provide a measure of customer service. In such a situation, customers are in effect being *starved* by the system. In output control, i.e. normally customer push systems, service may be measured by the occurrence or duration of queueing. Where queueing is required customer service may be measured by the average duration of queue or the number of items in the queue. Where queueing is not normally required, customer service may be measured by the probability that queueing will occur. In such situations customers are in effect being *blocked* by the system.

Customer service, whether in input or output control systems, has inventory cost implications e.g. costs of shortage, loss of trade, etc. Inventory and customer service costs are summarized in Exhibit 12.1, which also indicates their relevance in input and output control, continuous and intermittent flow situations. Costs will influence inventory management decisions; in fact inventory management decisions will generally represent a trade-off of costs of e.g. holding costs against change costs (see below).

INTERMITTENT FLOW/INPUT CONTROL

Most published treatments of inventory management deal with input control of intermittent flow systems, that is they deal with the management of inventory through manipulation of supply with the objective of satisfying a given output need or criteria. There are basically two approaches that might be adopted, namely:

(1) Fixed input level and quantity, and
(2) Fixed input interval.

The two approaches are compared in Exhibit 12.2. Fixed input level control relies upon the replenishment of stock by a given input quantity, actuated at a given inventory level. In other words inventory will fall to a re-order level whereupon replenishment is initiated or takes place. This approach is sometimes known as the 'max/min' or 'two bin' system. In contrast the fixed interval sys-

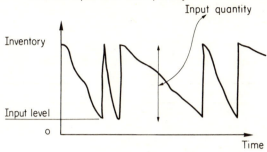

A Fixed input level and quantity

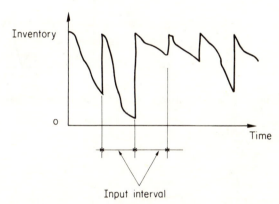

B Fixed input interval

Exhibit 12.2 Input control inventory—types of control

tem relies upon the replenishment of inventory at fixed intervals of time. The replenishment quantity in such situations is often determined such as to replenish inventory to a given maximum level. A replenishment of stock in input control intermittent flow systems might take place instantaneously or over a period of time. The stock level traces on Exhibit 12.2 rely upon instantaneous replenishment, i.e. replenishment of stock by a whole quantity at an instant in time. Exhibit 12.3 compares (A) intermittent input/instantaneous with (B) intermittent input with usage. The latter relies upon replenishment of stock intermittently yet over a period of time during which usage or output continues to occur.

Where demand is constant and replenishment is instantaneous or where replenishment lead time is known, the fixed input level approach will resemble the fixed input interval approach. Only where either replenishment lead time or demand is uncertain will the adoption of each approach lead in practice to different inventory behaviour (e.g. Exhibit 12.2 in which the diagrams show the

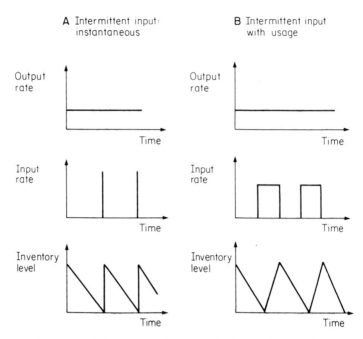

Exhibit 12.3 Input control inventory—types of input flow

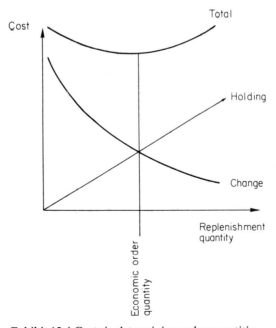

Exhibit 12.4 Costs in determining order quantities

effect of each policy against the same demand patterns). Virtually all inventory control quantitative models deal with intermittent flow and input control, i.e. batch ordering, and in most cases the objective is cost minimization, i.e. the minimization of the total of holding and inventory change costs (see Exhibit 12.4). Most such models are deterministic, i.e. they assume a known constant demand and either known input rate with no lead time or instantaneous input and known lead time. In such deterministic situations (which rarely if ever occur) there will be no need for provision of buffer or safety stocks. Such stocks will be provided only to protect against uncertain demand and/or lead time (see below). Probabilistic models are available as are models aimed at profit maximization, etc. Exhibit 12.5 provides a taxonomy of intermittent flow inventory models. The principal models are summarized below.

The use of a *queueing theory* approach might be of relevance in input control cases, particularly in demand push systems. In some cases there may exist some

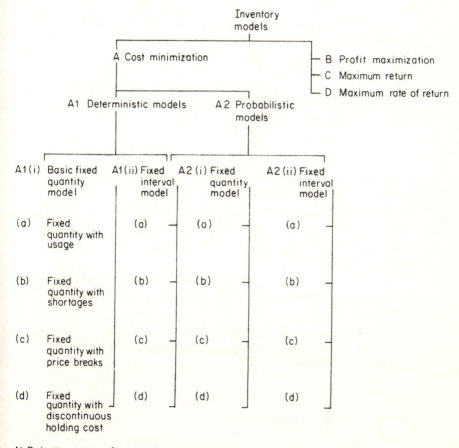

N.B Letters, etc. refer to text

Exhibit 12.5 Intermittent flow/input control, inventory models

opportunity for scheduling the arrival of customers in which case an arrival rate might be established such as to provide for efficient utilization of available resources.

A. Cost minimization[1]

A1. DETERMINISTIC MODELS, i.e. output rate (or demand or usage), holding and ordering costs *known*.

(i) *Basic fixed order quantity model* (i.e. input of complete batch, no shortages, no price or cost discontinuities).

$$EOQ = \sqrt{2C_s r/C_1}$$

where EOQ = Economic order quantity
C_s = Ordering cost/order
C_1 = Holding cost/item/unit time
r = Usage or output rate

(i)*a Fixed order quantity with usage* (i.e. input at known or constant rate, no shortages, no price or cost discontinuities).

$$EOQ = \sqrt{2C_s r/[C_1(1 - r/q)]}$$

where q = input rate

(i)*b Fixed order quantity with shortages* (i.e. input of complete batch, no price or cost discontinuities).

$$EOQ = \sqrt{2rC_s/C_1} \ \sqrt{(C_1 + C_2)/C_2}$$

where C_2 = shortage or stockout cost/item/unit time.

(i)*c Fixed order quantity with price breaks* (i.e. input of complete batch, no shortages but with discontinuous price discounts).
The technique involves calculation as in A1(i) for the various price ranges.

(i)*d Fixed order quantity with discontinuous holding costs* (i.e. input of complete batch, no shortages, no price breaks but with discontinuous holding cost).
The technique involves an iterative procedure similar to A(i)*c*.

N.B. Although rarely used, models are available similar to (i)*b*, *c* and *d*, for systems with usage.

(ii) *Fixed interval model* (alternatively known as order cycle systems) (in which orders, usually for different quantities are placed at fixed intervals).
e.g. For input of complete batch, no shortages and no price or cost discontinuity:

$$T = \sqrt{2C_s/rC_1}$$

where T = Optimum order interval
hence $EOQ = Tr$

N.B. Similar models are available for input with usage, price and cost breaks, shortages, etc.

A2. PROBABILISTIC MODELS[2]

In such models the assumption of known and constant demand is relaxed. Hence some stock must be held to protect against higher than expected demand. This degree of protection is usually based upon one of two criteria, i.e.:

(1) Set a safety or buffer stock to provide a required *customer service level* (i.e. a given probability of stock-outs), or
(2) Set buffer stock at a level which *minimizes the cost of shortages and the cost of carrying added inventory.*

Notice that in both cases the buffer or safety stock is protecting against higher than expected usage during the stock replenishment lead time thus variation in usage rate and replenishment lead time are critical (see Exhibit 9.3).

B. Profit maximization

C. Maximum return

D. Maximum rate of return

Other, somewhat more complicated inventory models are available aimed at, for example, the establishment of batch sizes which satisfy criteria other than minimum cost. Three such criteria are profit maximization, maximum return and maximum rate of return. Details on these approaches are available elsewhere.[3]

INTERMITTENT FLOW/OUTPUT CONTROL

The need for output control occurs mainly in respect of customer queues.

Fixed level and fixed interval approaches apply also in this context, in this case representing:

(1) depletion of stock by a fixed quantity when stock levels rise to a given output level, and
(2) depletion of stock by differing amounts at given intervals (comparable situations to those shown in Exhibit 12.2).

Furthermore intermittent output might occur instantaneously or with replenishment in the manner shown in Exhibit 12.6. Again the two approaches are similar to those described for input control systems.

Conventional inventory theory largely ignores the case of intermittent output control systems although there is an obvious resemblance with input control models. The two basic models shown in Exhibit 12.6 resemble the basic fixed quantity model (A1(i)) and the fixed quantity model with usage (A1(i) *a*). The basic fixed quantity, fixed interval deterministic, probabilistic models will apply in respect of intermittent flow output control models. The fixed order quantities with usage will correspond in this case to fixed order quantity with replenishment

142

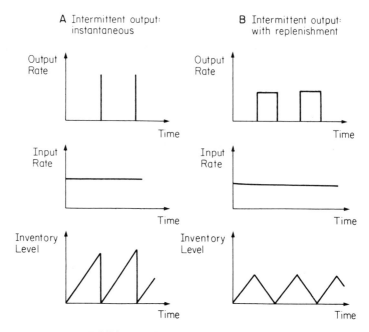

A Intermittent output:
instantaneous

B Intermittent output:
with replenishment

Output Rate / Time

Input Rate / Time

Inventory Level / Time

Exhibit 12.6 Output control inventories

but each of the other variations on the basic models, i.e. b, c and d of Exhibit 12.5 will not apply in the case of output control intermittent flow terms.

The use of a *queueing theory*[4] approach may be of relevance in customer push systems. For non-controllable inputs a knowledge of the arrival rate of customers and the service rate of the function will permit calculation of average queue length and waiting time. For a given maximum or average waiting time and a given arrival rate a required service time and service channel utilization (i.e. capacity and capacity utilization) can be calculated and hence such an approach is of relevance in the determination of capacity and the calculation of capacity utilization for given demand conditions.

CONTINUOUS FLOW/INPUT CONTROL

Conventional inventory control theory also largely ignores the case of continuous, as opposed to intermittent, input flow although again a queueing theory approach may have some relevance. Assuming input control as a means for the management of inventory to satisfy expected output needs, one problem for inventory management is the determination of an input rate (or average rate). Other problems will generally relate to the determination of average, minimum or safety inventory levels, and inventory capacity. Given deterministic output (or full control of output) and full control of input, input rate can be matched to output and inventory problems are obviated. Here, therefore we must deal with problems deriving from probabilistic output and/or incomplete control of input.

The problem can be considered to be one of matching two probability distributions (i.e. for input and output rates). A mismatch may give rise to:

(1) *Output starving* i.e. depletion of stock due to excess of demand over input (i.e. shortage, etc).

(2) (a) *Input blocking* i.e. insufficient space or capacity for inputs due to excess of input over output, or
   (b) *Excess stock holding*, if inventory capacity is not limited.

The required average inventory capacity will be influenced by input and output rate variability (mean input rate must equal mean output rate). The higher the variability, the greater the stock capacity required to accommodate short-term differences in input and output sales. Hence as a general rule of thumb the greater the possible short-term difference between rates (i.e. assuming symmetrical distributions, the upper end of one distribution minus the lower end of the other), the greater the stock capacity required. For a given (known or forecast) output rate distribution, inventory levels can be determined for alternative input rate distributions and vice versa. Simulation techniques will normally be employed.

CONTINUOUS FLOW/OUTPUT CONTROL

In this case precisely the same problems occur as in input control continuous input situations. The same techniques may be employed excepting that in this case the objective will be for a given (known or forecast) input rate distribution, either to establish inventory parameters for alternative output rate distributions or vice versa.

'NO CONTROL' SITUATIONS

The inventory control problem in such cases and where input and/or output are probabilistic and unpredictable, is simple, since there is no opportunity for control! There is no question of matching input and output rate distributions, yet in one respect the approach described above is appropriate. In this case simulation can be employed to determine requisite inventory capacity, capacity utilization, average stock, etc, for given or forecast input and output rates. In such cases the inventory management problem will often relate only to the determination of required capacity although in some cases a 'start-up' inventory might be provided (and perhaps in some cases maintained) to minimize input blocking or output starving.

SUMMARY OF CHAPTER 12

Procedures are available for the management of inventory through control of input or output, with continuous or intermittent flow.

144

Inventory costs (comprising holding and stock change costs) and costs associated with customer service will normally influence the choice and use of inventory management procedures.

Most quantitative inventory control procedures apply to the case of *intermittent flow/input control systems*, the principal procedures providing for either fixed input level and quantity, or fixed input interval replenishments.

Conventional inventory control theory largely ignores the case of *intermittent flow/output control*, although procedures similar to those relevant for input control can be developed.

Continuous flow/input control requires in effect the matching of two probability distributions, for input and output rates, to avoid *output starving* (depletion of stock due to excess demand), *input blocking* (insufficient space due to excess input), or *excess stock holding*. A similar situation applies in the case of *continuous flow/output control*.

REFERENCES

1. Wild, R., *Techniques of Production Management*, Holt, Rinehart and Winston, 1971, Chapter 12.
2. Zimmermann, H. J. and Sovereign, M. G., *Quantitative Models for Production Management*, Prentice Hall, 1974, Chapter 7.
3. Eilon, S., *Elements of Production Planning and Control*, Collier Macmillan, 1966.
4. Mitchell, G. H., *Operational Research*, English University Press, 1972.

Section 5

Operations Management Concepts, their Application and Development

This section reviews, refines and restates our concepts, and considers their applications. Some limitations of the concepts will be noted and possible further developments identified.

CHAPTER 13

The Concepts and their Applications

It is worth recalling that we set out to identify the *nature* of operating systems and the *nature* of operations management. In the preceding chapters we have developed a classification or taxonomy of systems, identified the principal operations management problem areas, discussed system characteristics, and looked at the nature of operations management in particular types of systems. In fact we have been dealing with four concepts or notions, as follows:

(1) The concept of *system structure*—the notion that operating systems can usefully be categorized according to their physical structure.
(2) The concept of *principal problem areas*—the notion of the existence of a relatively small number of principal distinguishing or characteristic operations management problem areas, i.e. capacity management, inventory management and activity scheduling.
(3) The concept of *system problem characteristics*—the notion that each structure will be characterized by certain types of problem in each of the three principal problem areas.
(4) The concept of the contrasting *nature of operations management* in different types of system—the notion that the job of operations management, i.e. the activities and the procedures employed, is in part influenced by the system structure and its characteristics.

In the light of our discussion in Sections 3 and 4 we can now further develop two of our concepts: those of system problem characteristics, and the nature of operations management. We noted that the nature of the operations manager's job was *in part* influenced by system problem characteristics (Chapter 2). We also noted in Section 3 that different strategies might be appropriate or indeed necessary in dealing with problems in different circumstances. Operations management's choice of strategy will be influenced not only by the problem characteristics but also by the objectives which operations management are pursuing. For example, in managing capacity certain approaches may be precluded because of the problem characteristics of the system. The use of output stocks, for example, may be impossible. However, some degree of choice will often remain, and such choice will be influenced by the relative priorities attached to customer service and resource productivity. Thus the nature and procedures

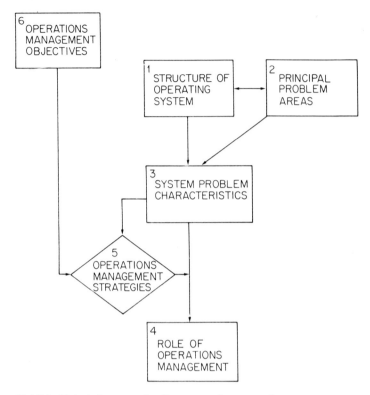

Exhibit 13.1 A framework of concepts for operations management

of operations management is influenced by both system problem characteristics and indirectly by operations management objectives.

We can now refine the framework used throughout this book. We can extend our diagram (Exhibit 2.1) showing the interrelationship of concepts in order to provide a clearer indication of the influence of objectives and the importance of operations management strategy. Exhibit 13.1 incorporates such modifications together with some revision of our earlier terminology (feedback loops are not shown). The term 'role of operations management' refers to the procedures and activities of operations management whilst the nature of the operating system is now clearly identified with system structure.

APPLICATION OF CONCEPTS

There is little point in developing concepts however elegant, unless they are of some value to us. Our emphasis throughout has been upon description and indeed this is the principal application of the ideas outlined above. The classification or taxonomy of systems by structure facilitates description, since using the seven basic structures we are able, using a common 'terminology', to model any operating system in any level of detail. Given the opportunity to examine a real system we can develop a representation or model as an aid to description. The characteristics of the system, strategies employed, and role of operations management can also be described using our terminology. Given a knowledge of the structure of the system and the objectives we can predict the problem characteristics of the system, the strategies which may be employed and the role of operations management without the need to examine the actual systems and a knowledge of objectives. The process of *description* therefore requires the use of concepts 1, 3, 4 and 5 above starting from a knowledge of the actual system, its environment and objectives. *Prediction* involves much the same sequence but originates from a given system structure and a knowledge of its environment and objectives, i.e. the prediction of 3, 4 and 5 above from a knowledge of 1 and 6 (Exhibit 13.2 illustrates).

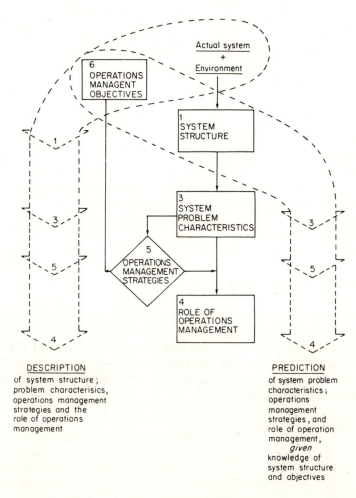

DESCRIPTION
of system structure;
problem characterisics,
operations management
strategies and the
role of operations
management

PREDICTION
of system problem
characteristics;
operations
management
strategies, and
role of operation
management,
given
knowledge of
system structure
and objectives

Exhibit 13.2 Use of concepts for description and prediction

Our concepts also facilitate *comparison*. Given the opportunity to observe actual systems, a comparison can be made of strategies and roles. Such comparison may aid selection; indeed, given the objectives required of a system and knowing the environment in which it will operate, alternative feasible systems might be identified, a comparison made of their characteristics, appropriate strategies and roles, in order that a choice can be made on the basis of desirability.

To some limited extent knowledge of this framework of concepts facilitates *explanation*. For example, the nature of the role of operations management in a given situation might be explained in terms of the prevailing objectives and system characteristics, the latter being explained in terms of the prevailing system structure. The explanatory process therefore in effect reverses the process of prediction. Whereas we might predict the managerial implications of alternative system structures we would perhaps choose to explain the nature of operations management by reference to the characteristics of the prevailing system structure.

In practical terms, this framework of concepts may enable operations managers to develop appropriate strategies, procedures, etc, to deal with new system structures, resulting perhaps from demand or product/service changes. A comparison of predicted strategies and roles for a given structure and objectives, with the strategies employed and the roles which exist, may reveal opportunities for improvement in present practices.

Summarizing, we suggest that the concepts and the framework outlined above might be of value in:

(1) *Describing* systems.
(2) *Predicting* the nature and characteristics of given system structures.
(3) *Comparing* systems and/or *selecting* from available feasible systems.
(4) *Explaining* the nature/characteristics of systems.

The brief case examples in Chapter 15 may help illuminate some of the points developed above.

SUMMARY OF CHAPTER 13

A framework has been identified to bring together six concepts, i.e. those of *system structure, principal problem areas, system problem characteristics*, the *role of operations management*, operations management *strategies*, and *objectives*.

This framework of concepts is of value in *describing* systems, *predicting* their nature and characteristics, *comparing* and *selecting* systems, and *explaining* their nature and characteristics.

Chapter 14

System Description

A descriptive treatment of operations management has been developed. Some concepts have been identified which might be of value in attempting to understand and describe operating systems and the nature of the management of such systems. Many simplifying assumptions have been made. Some of these will now be tackled as we try to indicate how our basic concepts might be employed in practice.

Two important complicating factors must now be considered if we are to see how these concepts might be used. In Chapter 4 the distinction was made between *single* and *multiple channel systems*. It was argued that in many cases the simple structures can accurately represent multiple channel systems where input resources are treated similarly in respect of the provision or absence of stock. In other cases two or more basic structures will be required to represent systems. In Chapter 4, in discussing hierarchies of systems we described a launderette as two sequential sub-systems, noting that given sufficient detail all systems might be considered to consist of several others arranged in *series and/or in parallel*.

Our objective in this chapter is to suggest how rather more complex systems might be described using our seven basic system structures. The final chapter will then look at some uses and possible developments of our concepts.

MULTI-CHANNEL SYSTEMS

Operating systems will normally require a variety of physical resources, e.g. materials, machines, and labour. If all such resources are treated similarly, i.e. if they are all stocked or alternatively if none of them are stocked, we can for our descriptive purposes consider the system as having only one resource input channel (plus the customer channel in the case of push systems). If some resources are stocked and others not, this approach cannot be adopted. In such cases we must consider the system to have two or more structures (see Exhibit 4.11). This complication in the description of systems is of particular relevance in the examination of the capacity management problem.

When considering the available strategies for the management of capacity, i.e. that of providing for capacity adjustments, and/or minimizing the need for such adjustment, and the means available for the implementation of such strategies,

152

it may be necessary to employ a different approach for each resource used by the system.

In other words when examining the strategies available (see Exhibit 7.4), it may be necessary to consider each channel as a system and to employ a mixed strategy for capacity management for the system as a whole.

From all other respects, i.e. inventory management and activity scheduling, the configuration of the system has little direct effect upon the relevance of the characteristics, problems, etc discussed in previous chapters.

Exhibit 14.1 shows a multi-channel system and suggests the approach to capacity management which might be appropriate.

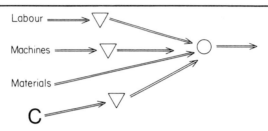

Capacity Planning Strategy: 2bI + 1 (See Exhibit 7.4)

The use of this structure suggests that there is sufficient predictability of the nature of demand to permit the stocking of non-consumed resources. Uncertainty as to the nature of the materials required for future orders may account for the absence of stocks. Customer queueing and perhaps loss of trade will be employed to minimize the need for capacity adjustment, however in addition strategy 1 may be employed for all resources, especially if the customer service objective is of importance.

i.e. (1) *Labour and Machines* (i.e. as structure SQO)
      Strategy 2bI
      with
      Strategy 1

    (2) *Materials*   (i.e. as structure DQO)
      Strategy 1 may be necessary if there is any uncertainty of material requirement for each order

Exhibit 14.1 Capacity management for a complex system

SERIES AND PARALLEL SYSTEMS

By employing an appropriate level of detail in describing a system, any operating system can be represented by one of our seven basic system types. We have argued that an understanding of the characteristics of each basic system type will be of relevance in considering systems irrespective of their hierarchical level; for example, that a knowledge of the characteristics of system structure SOD will be of relevance whether we consider a retail counter or a departmental store. However, in order to be able to adopt this approach we must be able to identify where, for example, the retail counter as a system fits into the departmental store as a system. In other words, as well as being able to recognize the latter as a system of structure SOD, we must also be able to see it as a complex of sub-systems and recognize the structure of these sub-systems. We must be in a position to describe any level of system in any level of detail in the manner of Exhibit 4.2. We must be able to build complex systems using the seven basic system structures as our building blocks. This will enable us to take an operating system as we find it, see it as a simple system, and have some idea of the overall operations management problems involved, as well as locate parts or departments of the whole, identify their structure and characteristics and have some understanding of the problems involved in their operation. This section therefore examines how the seven basic system structures might be pieced together, in order to provide a detailed description of more complex systems.

Series/customer pull systems

Consider firstly the case of 'complex' systems consisting of a series of functions together aiming to satisfy a customer demand or pull; in other words we are considering series arrangements of structures SOS, DOS, SOD and DOD.

Exhibit 14.2 shows one such system, consisting of two sub-systems. Were we to consider this as one system it would be seen as structure SOS, i.e. 'function from stock to stock to customer' as in Exhibit 14.3. If, however, it is taken in the detail shown here, it is either SOS→DOS or SOD→SOS, as shown in Exhibit 14.4.

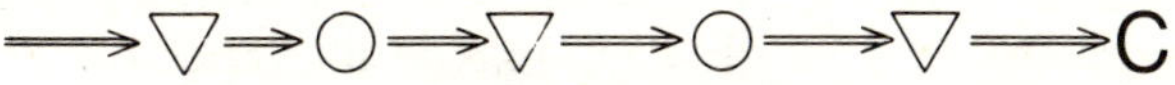

Exhibit 14.2 Two sub-systems in series

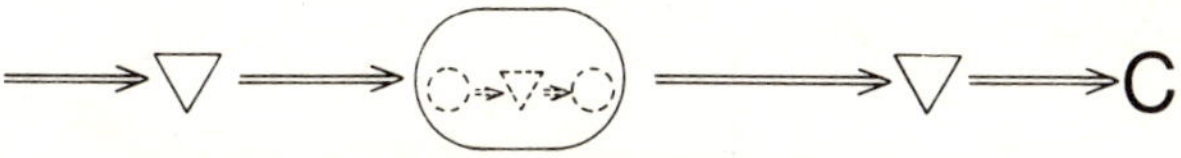

Exhibit 14.3 Exhibit 14.2 as a simple system (SOS)

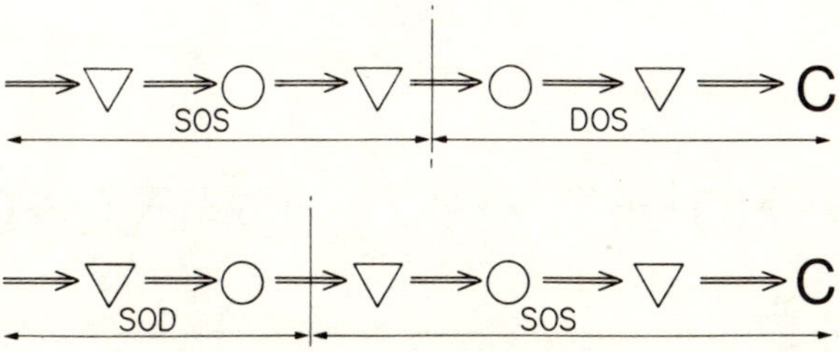

Exhibit 14.4 Alternative descriptions of two sub-systems
in series

154

The choice of description depends upon our views of the purpose of the intermediate stock. If the intermediate stock belongs to the first function, we would see the arrangement as SOS→DOS. If it belongs to the second function we would see the system as SOD→SOS. 'Belonging' in this context must relate to the use and purpose of the stock. We can consider the intermediate stock to 'belong' to the first function if it can serve the purposes we have identified for system output stocks, i.e. that of insulating the function against fluctuation in demand level and thus permitting the adoption of a capacity management strategy based on reducing the need for adjustments in system capacity. If we see this intermediate stock as satisfying the purpose of system input stock, then logically we would consider it to belong to the second sub-system.

The nature of the management of the intermediate stock will be much the same irrespective of whether the stock is considered to belong to the first or the second sub-system. In either case stock input will be controlled. Similarly, the nature of the activity scheduling problems in the system are largely unaffected by our views as to the 'location' of this intermediate stock. Whether the system is seen to be SOS→DOS, or SOD→SOS is therefore mainly a function of how we view the nature of the capacity management problem in the system.

In fact the intermediate stock serves two purposes. It insulates the first function from those fluctuations in demand levels which are evident in the input to the second function *and* it serves as an input resource stock to the second function. We can therefore consider this stock to serve, or 'belong', to both sub-systems, and can therefore view this series system as being SOS→SOS. In other words, in all important respects the characteristics and the nature of the management problems for each of the two sub-systems are as described for system structure SOS in the preceding chapters. Exhibit 14.5(A) illustrates this interpretation.

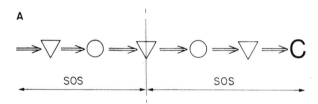

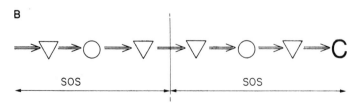

Exhibit 14.5 Two sub-systems in series

Exhibit 14.5(B) shows the alternative configuration for an SOS→SOS system. This latter case will rarely occur. In fact, we can for our purposes consider such a pairing to be impractical in this and all other types of complex system configurations.

Most sequential configurations of systems can be treated in a similar manner. For example, Exhibit 14.6 shows a four-stage system, which can be considered as SOS→SOD→DOS→SOD.

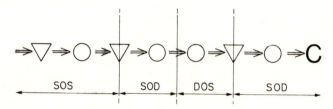

Exhibit 14.6 Four functions in series

Series/customer push systems

A two sub-system arrangement is shown in Exhibit 14.7. If we consider this as one system it would be described as having a structure SQO (Exhibit 14.8). Taken in the detail drawn here it can be described as SQO→SQO (Exhibit 14.9). Notice that unlike the cases above there is no problem in dealing with intermediate stocks since these can only be considered as systems input stocks. Furthermore, all pairings of systems are feasible.

Exhibit 14.7 Two customer push sub-systems

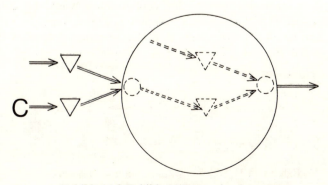

Exhibit 14.8 Exhibit 14.7 as a simple system

Most sequential configurations can be treated in a similar manner. For example, Exhibit 14.10 shows a three-stage SCO→SQO→DQO system.

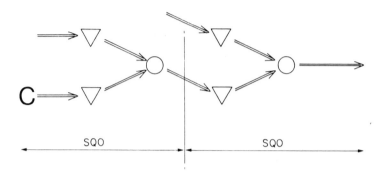

Exhibit 14.9 Series of two customer push systems

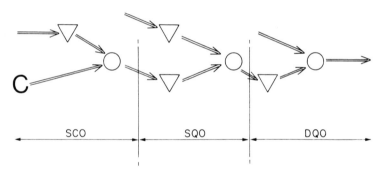

Exhibit 14.10 Series of three customer push systems

Series/customer pull and push systems

Exhibit 14.11 shows a two sub-system arrangement involving both customer pull and push. As a single system this configuration would be considered as having structure SQO, whilst as a complex system it must be viewed as SOS→SQO (Exhibit 14.12). We should note that in mixed systems of this type where the customer himself is input the pull sub-system must precede the push system.

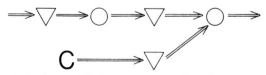

Exhibit 14.11 Series customer pull and customer push systems

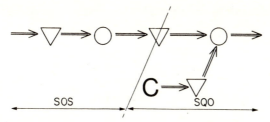

Exhibit 14.12 Exhibit 14.11 as a complex system

Parallel/customer pull systems

Exhibit 14.13 shows a simple three sub-system parallel arrangement. As a single system it is structure SOS with two input channels. As three sub-systems it must have the following configuration:

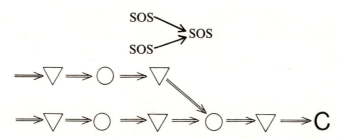

Exhibit 14.13 Sub-system in parallel

A similar approach can be employed in describing parallel arrangements of a larger number of sub-systems, although in some cases in order to provide an accurate description it will be necessary to consider some sub-systems as having multiple input channels in the manner outlined above.

For example, Exhibit 14.14 shows a parallel arrangement incorporating three sub-systems. There are three inputs to the final function, but one of those does

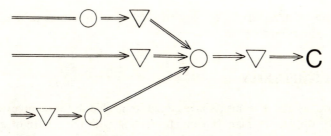

Exhibit 14.14 Sub-systems in parallel

not provide for input stock. Hence, it is not appropriate to consider this final sub-system to have structure SOS. In this case we must use two basic structures to describe the final sub-system, therefore the overall system might be described as follows:

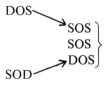

Parallel/customer push systems

Parallel arrangements in which customers are input will rarely, if ever, exist in practice, since it is improbable that any function will depend upon the simultaneous processing of two people.

Parallel/customer pull and customer push

Exhibit 14.15 shows a two sub-system configuration of this type. As a single system we might consider this to be adequately described as structure SQO in

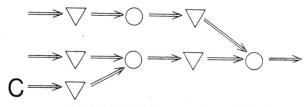

Exhibit 14.15 Sub-systems in parallel

this case with two input resource channels. In the detail employed here we would describe the system as:

Again it may be necessary to employ two or more structures to describe one sub-system, as outlined above.

SYSTEM COMPLEXITY

Inevitably, the above seems to have added somewhat to the complexity of our treatment of operations management. In dealing with the *reality* of operating systems (i.e. multiple channels, etc) it has been necessary to resort to the concept of 'complex' systems. This complexity, however, is more apparent than real. In fact, we are using only the one basic system 'building block' shown in Exhibit 14.16(A) in describing systems. Depending upon the location of the customer and

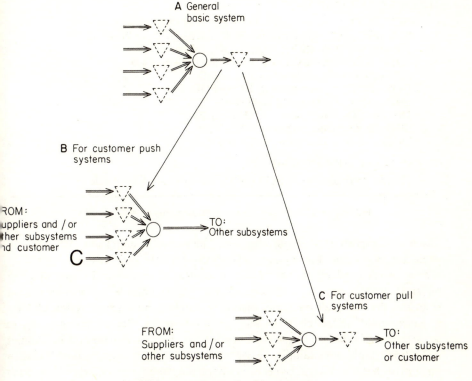

Exhibit 14.16 Generalized basic system

the location of stocks this 'generalized basic' system will represent any one of our seven basic structures. Exhibit 14.16(B) and (C) illustrate.

We hinted in Chapter 4 that it might be possible to represent any system using fewer than seven basic structures. The use of this generalized basic system structure provides one means to achieve this. Clearly, we can consider customer input as one possible input channel; however we should not forget that it differs from other input channels in that it affords less opportunity for management control. It would therefore be an oversimplification to treat that channel as any other input resource. This push/pull distinction is important and should not be overlooked when describing complex systems. This danger apart there is perhaps some benefit in considering all systems to constitute arrangements of the generalized basic system shown in Exhibit 14.16(A). This generalized basic system is the lowest common denominator in our description of operating systems. Using this system subject to the rules developed in this chapter and summarized in Exhibit 14.17 we should be able to model or describe any real system. (Furthermore, it permits us to accommodate the fact that in some complex systems a customer push system may precede a customer pull system; thus there may be more than one customer in the system. For example, a bakery produces bread

(1) SERIES ARRANGEMENTS

(SCO, DQO, or SQO)→(SOS, DOS, SOD, or DOD) are infeasible
SOS→SOS; SOS→SOD; DOS→SOS; DOS→SOD; SOS→SCO;
SOS→SQO; DOS→SCO; DOS→SQO; are improbable
SOD→SOS; SOS→DOS; are interpreted as SOS→SOS
SOS→DOD; is interpreted as SOS→SOD
DOS→DOD; is interpreted as DOS→SOD
SOD→SCO; is interpreted as SOS→SCO
DOD→SCO; is interpreted as DOS→SCO
SOD→SQO; is interpreted as SOS→SQO
DOD→SQO; is interpreted as DOS→SQO

(2) PARALLEL ARRANGEMENTS

Parallel arrangements of SCO, DQO, and SQO are improbable

Exhibit 14.17 Limitations on arrangements of sub-systems

which is transported to the shop for sale. The second system transports bread—its customer—to the third system—the shop—for sale to its customer.)

It is worth recalling that we have concentrated on physical flows, and the physical characteristics of systems. Were we to consider information as a resource, the customer push/pull distinction and its influence on system characteristics would be less clear. In fact, we might argue that all systems require some customer input—a physical input and/or an information input. (A customer order book could, for example, be considered as an input resource stock.) If such an approach is taken it will be necessary to employ a somewhat different approach to describing or modelling system types; however, it will still be important for us to distinguish customer controlled channels from those controlled by operations management.

SUMMARY OF CHAPTER 14

In *multi-channel systems* it may be necessary to employ a different capacity management strategy for each channel.

Large systems may be described in detail by using *series and/or parallel arrangement* of the seven basic structures. Such *complex systems* may alternatively be represented by configurations of a generalized basic system structure.

Chapter 15

Case Examples

The following brief case examples may help clarify some of the points developed in Chapters 13 and 14 as well as illustrating some of the limitations of our concepts. We shall look firstly at drinking, secondly at eating, and finally at a public section system.

CASE 1: AN ENGLISH PUB

The typical English country pub provides both drinks and food at certain times during the day. Whilst their primary purpose is to supply alcoholic drinks for consumption on the premises during prescribed hours, most pubs now provide snacks and/or restaurant-type meals at lunch times in order to attract more custom. We shall consider one such establishment, which for our purposes will be called the 'Stag and Huntsman'. It is a non-residential pub providing bar snacks (e.g. soup, sandwiches, etc) at lunch times.

As a simple system the Stag and Huntsman may be *described* as a SOD supply system (Exhibit 15.1). This description ignores all but the supply aspect. For example, tables, chairs, etc, are provided for use by customers whilst con-

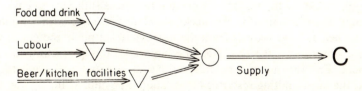

Exhibit 15.1 An English pub as a simple system

suming food and/or drinks. Thus, there is a service system operating in parallel with the supply system, both having the same customers. In more detail we might identify the preparation of food as a separate system, thus Exhibit 15.1 becomes the more complex system shown in Exhibit 15.2. It will be seen that the publican devotes himself to work at the bar whilst his wife operates the manufacturing system. A stock of cold food (e.g. sandwiches) is built up prior to the lunch period. Hot food (e.g. soup) is made only to order, indeed from about 1 pm onwards the stock is depleted and all food is made to order, often with substantial delays.

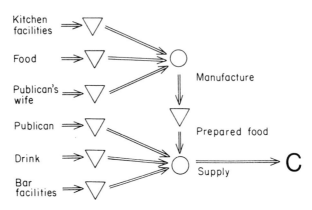

Exhibit 15.2 An English pub as a complex system

During the evenings the Stag and Huntsman provides for drinkers only, and both the publican and his wife are required to meet demand at the bar.

The *characteristics* of the system ensure that management of the pub is relatively straightforward. There are few scheduling problems, inventory management is uncomplicated, and there are few available capacity management strategies. The publican uses his own job experience directly to forecast future demand. He knows that the fluctuations in demand are regular—the highest demand rate during each day, week and year are, respectively, approaching closing time, the weekend, the summer. Through his weekly inventory decisions he ensures stocks of the major consumable resources sufficient to cover anticipated maximum demand for the week. He claims to cope with unpredictable demand, such as large crowds from the nearby cricket pitch and coach parties, by being unpleasant, thereby discouraging trade and protecting his regular customers. For the most part, fluctuations in demand are predictable and modest, the most limiting resource being labour. To cope with 'minor' fluctuations in demand he operates a double-manning system for himself, and in addition calls upon the services of his wife on a flexible work shift pattern with overtime for both is necessary (for cleaning up in summer, etc). During times of over-capacity he can perform maintenance and other intermittent functions. Should there be under-capacity at lunch times, the stock of prepared food disappears. Should under-capacity increase, the food function is eliminated altogether.

From a capacity management point of view, the Stag and Huntsman is managed through a mixed strategy, roughly as follows:

Drinks	*Food*
Strategy 2bI	Strategy 2bII
(with loss of trade in extreme cases, and customer waiting as a common phenomena)	plus
	Strategy 2bI
	(with loss of trade and customer waiting)
plus	plus
Strategy 1	Strategy 1
(implemented by varying working hours, deferring non-essential work; transferring labour from and to the 'food system', etc)	(implemented by varying working hours, deferring non-essential work, etc)

Clearly the provision of drinks attracts far greater priority than the manufacture of meals, hence in extreme cases the latter system is forfeited in order that the pub might continue to provide acceptable customer service in respect of its principal activity. In much the same way and for much the same reason, customers may be lost in order to maintain service to the 'regulars' upon whom the establishment depends.

Whilst other system structures may be feasible to some limited extent, they are undesirable. Drinks could be supplied to stock but this is unlikely to be done except for those 'regular' customers who arrive at precisely the same time on given days and always require the same drink. All food could be preprepared but this would necessitate a more limited menu, and the use of special storage facilities. Such a system is employed in many city pubs as the only practical means to meet very high, short-term yet commercially important lunchtime demand for food.

CASE 2: THE TAKE-AWAY AND THE RESTAURANT

The take-away meal shop or 'house' is one of the more prominent aspects of modern living. Although most noticeable in the USA where hamburger, pizza and doughnut houses abound, it is a significant feature elsewhere—for example the traditional British fish and chip shop. All such establishments seek primarily to provide rapid 'no frills' service for customers who prefer not to wait. (Significantly they are less a part of French eating habits.)

We will take a typical hamburger house as our first case example. At most times of the day customers, who will usually arrive by car must queue briefly at one of the several order/collection/cash points. Orders placed here will be satisfied virtually immediately since a small stock of preprepared food is maintained and located between the kitchen area and the service area. Customers pay for the goods and take them away to eat usually in their car or at home. Considering one channel only, the hamburger house could be described at a simple level as a system type SOS with the function of manufacture. In more detail we would

164

describe the system as comprising two sequential sub-systems, i.e. SOS→SOD with the functions of manufacture and supply (Exhibits 15.3 and 15.4).

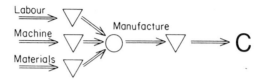

Exhibit 15.3 A hamburger house as a simple system

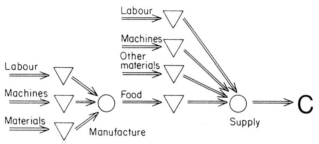

Exhibit 15.4 A hamburger house as two sub-systems in series

In a typical restaurant, in which customers are required to make prior table reservations, the simple system *description* in Exhibit 15.5 will apply. In all but exceptional cases, customers must wait during the period required for the preparation of their meal, before being served by the system. All resources required in this service system are stocked excepting meals themselves which are the

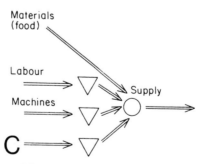

Exhibit 15.5 A restaurant as a simple system

output of a prior manufacturing system. Hence at a more detailed level, a typical restaurant might be described as two sequential systems comprising manufacture and supply as shown in Exhibit 15.6.

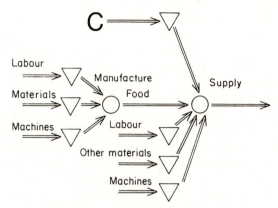

Exhibit 15.6 A restaurant as two sub-systems in
series

These system models provide one basis for *comparison*. The principal structural difference is the customer pull nature of the hamburger house system as
compared to the customer push system (or part of the system) for the restaurant.
We now also have a basis for *prediction*. It is clear from the different systems
structure in these two cases that both the system characteristics and the nature
of the role of operations management will differ. The fundamental structure
differences result from the differences in the situation external to the two systems.
The provision of output goods stocks in the hamburger shop is permitted by the
predictability of the nature of demand, which in turn is the result of the existence
of a limited menu (no more than eight main items are available on the hamburger house menu). In contrast, the menu in the restaurant is of necessity
somewhat longer, hence meals cannot be prepared prior to the arrival of the
customer and some customer waiting is therefore inevitable. Whilst both systems
exist to feed customers, their different approaches to this task reflects their
different attitudes to the objective of customer service. In the case of the hamburger house, speed of service is of paramount importance and less importance
is attached to product choice. In the case of a restaurant, considerably more
importance is attached to the menu, it being expected that given suitable products
customers will be prepared to wait some time to be served. Without the limited
menu a hamburger house would of necessity operate in a different manner. The
Chinese take-away, by virtue of the need to offer a considerable choice of meals,
must require customer queueing to a far greater extent than is evident in a
hamburger house. It operates with a system structure type SOD. American
ice-cream parlours would operate in the same manner as take-aways. In both
cases, however, the preparation or manufacturing time is sufficient to permit
relatively short customer queues and therefore management has in practice the
opportunity of choosing to operate in the manner of a take-away or in the manner
of a restaurant. In practice many offer both arrangements. Such a choice is not
available to the average restaurant owner simply because of the type of meals
he chooses to offer. In fact, given the predictability of the nature of the demand,

four systems structures are feasible for the take-away (SOS, DOS, SOD, DOD) but only two (SOS and SOD) are practical. In contrast, because of the nature of demand, only two structures (SOD and DOD) are feasible for the restaurant and only one, i.e. SOD, is in fact desirable.

The nature of the structure of the two systems would lead us to predict different characteristics for each in respect of the three principal problem areas. This is particularly evident in management of capacity. The existence of output goods stocks in the case of Exhibit 15.3 affords a strategy of minimizing the need for changes in system capacity (strategy 2bII). However, since stock turnover must be rapid to avoid product deterioration, this capacity management strategy alone will be insufficient. Furthermore since demand level will fluctuate there will clearly be a need to vary system capacity to ensure reasonably high resource utilization throughout the day. There is little opportunity for transferring resources within the system since both the supply and the manufacturing function are affected in the same way by demand level changes. Increased demand will of course lead to increased customer waiting time, whereas reduction in demand will lead to increased stocks if capacity is not reduced. There will be an upper limit on capacity, hence some customer queueing will be required and customers will be lost. Additionally, a policy of adjusting capacity to match major demand level changes will be adopted through the use of different working shifts, part time workers, etc. In the case of the restaurant, minimizing the need for capacity adjustments is afforded by the opportunity of rejecting some customers, i.e. accepting only a given number of bookings each day (strategy 2bI). Reduction in demand levels will in the short term give rise to an underutilization of capacity, but predictable demand level changes will to some extent be matched by the use of part time workers, different working shifts, etc. Most customers will queue for some time through the restaurant system and an increase in queueing time will have less effect upon their satisfaction than in the case of the hamburger house. Even with normal reservations systems increased waiting time is likely at periods of high demand. There is, of course, a tendency in such systems to try to 'occupy' the customer during this waiting period. For example, as in the case of many pub restaurants in the UK, customers may be required to place their orders for food whilst sitting in the bar where they can of course be served drinks on virtually a 'no-wait' basis. This has the added benefit of permitting higher utilization of the restaurant resources, since fewer tables will be occupied by customers waiting for their meals. The tendency is to introduce further operations in order to occupy the period of time which would otherwise be spent waiting for the preperation of the product.

Both systems are sensitive to demand level fluctuations. The hamburger house must seek to accommodate demand increases and reductions. Increased demand levels will, unless accommodated by the system give rise very quickly to loss of trade and increased customer dissatisfaction whilst reduction in demand may result in wasted output. The restaurant is less sensitive to temporary demand increases since customers are more likely to expect to wait for service. Demand reductions will give rise to resource underutilization but not loss of

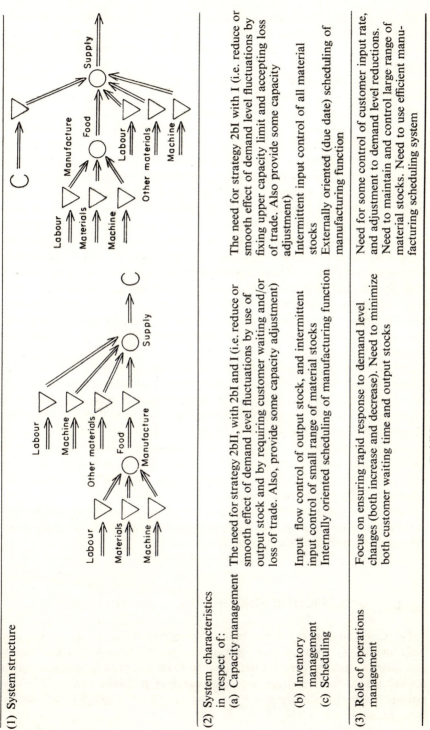

System	Hamburger House	Restaurant
(1) System structure		
(2) System characteristics in respect of:		
(a) Capacity management	The need for strategy 2bII, with 2bI and I (i.e. reduce or smooth effect of demand level fluctuations by use of output stock and by requiring customer waiting and/or loss of trade. Also, provide some capacity adjustment)	The need for strategy 2bI with I (i.e. reduce or smooth effect of demand level fluctuations by fixing upper capacity limit and accepting loss of trade. Also provide some capacity adjustment)
(b) Inventory management	Input flow control of output stock, and intermittent input control of small range of material stocks	Intermittent input control of all material stocks
(c) Scheduling	Internally oriented scheduling of manufacturing function	Externally oriented (due date) scheduling of manufacturing function
(3) Role of operations management	Focus on ensuring rapid response to demand level changes (both increase and decrease). Need to minimize both customer waiting time and output stocks	Need for some control of customer input rate, and adjustment to demand level reductions. Need to maintain and control large range of material stocks. Need to use efficient manufacturing scheduling system

Exhibit 15.7 Comparison of systems

products. The hamburger house must rely upon a fairly efficient communication system, through both inventory and scheduling procedures, in order to ensure rapid adjustment to demand level changes. This is less important in the restaurant system.

The management of inventories in each system will involve input control. The output stocks in the hamburger house will be low and turnover will be rapid. The flow will be virtually continuous, hence the objective will be to match input and output flow rates. The manufacturing system will seek to use its resources so as to ensure that a minimum stock level is held for each major product. Because of the close proximity of the supply and the manufacturing functions, those responsible for manufacturing will be able to observe inventory levels and adjust their operation accordingly. In addition to changes in aggregate demand level during the day, product mix changes may also occur. Some knowledge of likely changes will assist the management of the manufacturing function, although in practice those involved in manufacture will respond to the state of inventory rather than the requests of individual customers. The management of inventory materials will also be exercised through input control, although in this case input will be intermittent. A somewhat simpler inventory management situation occurs in the restaurant. The major problems will concern both materials employed in manufacture and supply. In both cases replenishment will be on an intermittent basis with control exercised over input.

Scheduling within the hamburger house is relatively simple and whilst output inventories exist there will be little direct communication between personnel throughout the process. Since those involved can observe activities at all stages in the process, the scheduling of manufacturing will result from observations of the state of the output inventory, i.e. scheduling is internally oriented. In the rare cases when inventories are eliminated customer orders must be communicated directly to manufacture. This is the normal situation for the restaurant where the scheduling of manufacture is externally oriented, being affected directly by customer orders.

Exhibit 15.7 summarizes this comparison of the two systems and indicates the manner in which the role of operations management might be explained by the system characteristics and structure.

CASE 3: A PUBLIC HOUSING BUILDING PROGRAMME

This case is somewhat more complex in being concerned with a very long time-scale and largely 'professional' operation. Physical flows do not predominate and various other factors limit the extent to which our principles can usefully be applied for descriptive, predictive or explanatory purposes. We shall however attempt an analysis utilizing our concepts in order to highlight some of their limitations.

System description

The programme exists to provide rented residential accommodation for residents of a large city. It consists in effect of a 'pipeline' comprising six sequential sub-systems together necessary to create houses for people from a lengthy waiting list. Exhibit 15.8 shows these six sub-systems as part of the total system and each sub-system is outlined below:

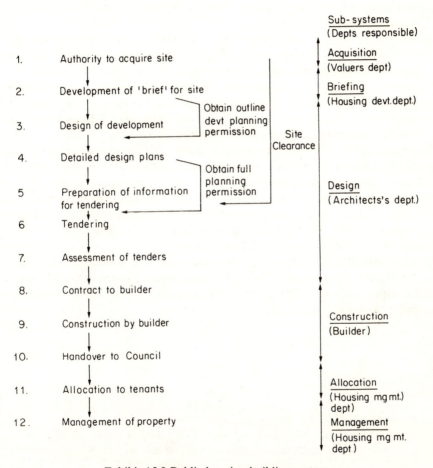

Exhibit 15.8 Public housing building system

(1) ACQUISITION   The Valuers Department, in drawing on its own personnel to select and acquire sites, has an input stock of labour, as well as the materials, equipment, and building in which to carry out that job. There is no input stock of sites; however, there is an output stock of sites prior to tendering. Depending upon whether the emphasis is placed upon the pool of surveyors or on the actual things to be processed, the sub-system has structure SOS or DOS.

(2) BRIEFING   The Housing Development Department draws on its personnel, and may or may not have an input stock of sites. The brief goes straight to the Architects Department or private architects. The sub-system therefore has either structure SOD or DOD.

(3) DESIGN   The Architects Department has a pool of architects, engineers and quantity surveyors, but briefs are not usually supplied to the Department unless it wishes to take them on. Normally no output stocks of completed sets of drawings are held hence the system can be considered, at a simple level as SOD or DOD.

(4) CONSTRUCTION   On being given the job, the builder arranges the drawings to suit his own approach and at the same time hires plant and labour and orders materials. In many cases the only thing the builder has in stock is his own (or his site manager's) expertise. Although in detail the building process may involve fabrication and assembly with stocks of materials on site, overall the sub-system has structure DOD, as the builder hands over the building as soon as it is complete.

(5) ALLOCATION   Whilst there is a queue of customers, i.e. potential tenants, they are not processed or 'treated' by the system at this stage, but simply allocated to dwellings. As this allocation is done as soon as the buildings (or indeed phased parts of buildings on large sites) are handed over by the builder, there are no input or output stocks other than the personnel involved in the Housing Management Department. The sub-system can therefore be seen either as SOD or DOD.

(6) MANAGEMENT   Here the customer enters the overall system and is 'treated' by the system, i.e. is provided with accommodation. There is no input stock of dwellings, but there is a queue of customers, hence the sub-system has structure DQO.

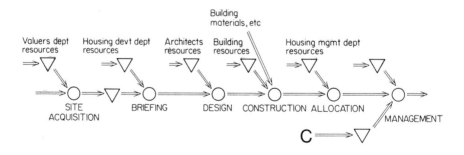

Exhibit 15.9 Public housing building system model

Arranging these systems sequentially the total system of Exhibit 15.9 is obtained. Alternatively the entire arrangement can be described as a single system which, taking account only of the principal physical resources, would have structure DQO (Exhibit 15.10).

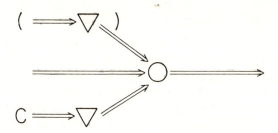

Exhibit 15.10 The public housing building programme as a simple system

This building 'pipeline' is designed to meet a target of approximately 5,000 housing starts per annum (i.e. 5,000 dwellings commenced on sites within each year). The lead time to that point varies between 3 and 5 years depending mainly on site acquisition and planning difficulties. The time of site acquisition is clearly important. The site must be clear by the time the design is ready for tendering, but early investment changes the nature of the planning permission, the other vital aspect of the process. Early investment, however, incurs interest charges, which, while being amortized over a projected building life of 60 years and therefore in the long run the same in total, is embarrassing to the extent that charges are being incurred when no use can be made of the site. Obtaining planning consents involves long procedures. These, combined with application for subsidies, etc, introduce problems which can tie up designers on administrative matters. To cope with these difficulties, and to meet the requirement of 5,000 starts per year, the 'pipeline' is approximately 10 years long. Approximately 58,000 dwellings with a total contract value of £850m to be built on approximately 600 sites are in the pipeline at any time, involving capital expenditure of about £120m per annum on new housing plus £20m for land acquisition.

When the Valuers Department has received authority to acquire a site, the Housing Development Department work up a brief with the valuer and the architect. The brief then either goes to a private architect, or to the Architects Department, the latter being staffed to a level just above what is perceived as the potential 'trough' level of work, peaks load being subcontracted to private architects. Within the Architects Department the briefs are allocated to three identical divisions, who in turn allocate it to one of their five identical groups.

Once a group has received a brief it is allocated by the Group Leader to a job architect, who may receive assistance from junior architects, depending on the size of the job and the stage it has reached. The job architect will head a 'job team' with mechanical, electrical, structural and civil engineers, a landscape architect and a quantity surveyor. These will be nominated from groups in their respective departments/divisions, and will be doing anything between 2 and 30 jobs.

System characteristics and management

In looking at the characteristics of this system we shall concentrate upon the Architects Department.

Clearly there are considerable payoffs in being able to balance *capacity* with demand. However, even with such a pipeline, the difficulties in smoothing the flow of sites acquisition means that demand will fluctuate. Of the two basic strategies (adjustment in systems capacity; and elimination or reduction in the need for adjustment in system capacity), the former approach is perhaps the more desirable as the second will tend to involve certain penalties. In practice, the Architects Department normally employs the first strategy, turning to the second with reluctance because of the nature of the professional work involved. In seeking to adjust capacity four possible tactics exist (see Exhibit 7.2). The first is the subcontract/sack tactic. The Architects Department tries, by setting staffing levels just above estimated potential trough levels, to ensure that this tactic is normally implemented in the form of subcontracting peaks to private architects. The second tactic is the buy in/do it yourself tactic. This, because of the ethos of the architectural profession, tends to mean the 'one-off' or 'proto-type' approach to in-house designs, with pressure to buy-in builders' systems when demand increases. However, even when demand is low, there is pressure to maintain the normal approach through the use of the Architects Department's own 'Preferred Dwelling Plans' (PDPs)—an attempt to improve design time by creating a component stock. The third tactic is to transfer staff to areas of pressure. Such an approach tends to prove difficult, mainly because of reluctance to move. The final tactic is to improve scheduling, which tends to be seen as the area with the greatest promise. The department maintains a planning network for a whole job, from which 'Plans of Work' are devised to assist job architects in running their jobs, to ensure that due dates are achieved.

The second strategy offers four tactics for coping with demand increases, two of which can also be used for dealing with decreases in demand (Exhibit 7.3). The provision of excess capacity is unacceptable especially in times of economic stringency. The creation or use of output stocks is sometimes advocated particularly as a means to avoid transfer or loss of staff. The difficulty here is that creating an output stock of completed drawings during periods of low demand may be futile, as they may be out of date by the time they come to be used. The 'loss of trade' tactic is necessarily accepted in the policy of staffing to just above estimated trough level—hence the occasional use of private architects. Finally there is the possibility of queueing, but there is a reluctance to adopt this approach because of the effect on interest charges on unused sites which, though unimportant in terms of total building life, is politically awkward in the short term.

Operations *scheduling* has been the tactic on which the Architects Department has placed most emphasis in dealing with demand fluctuations. Clearly the need to produce 5,000 starts per year provides an external due date schedule. Man day allowances for each stage of the design process are then calculated by dividing the architect's scale fee for the stage, minus overheads and contingency reserves, by the current man day cost for that section of the department. This gives the number of man days that can be spent between the due dates. If these exceed the number of calendar days, then more than one person must be allocated to the job, and if they are less, then the person assigned to the job has spare capacity.

With the exception of sites, there is little by way of on-line system inventory. *Inventory* control of convertible inputs—materials and equipment—is a fairly straightforward stock replenishment problem based on control of flow. Provision of offices is slightly more complicated but, given the decision to limit capacity, does not pose inherent problems. Inventory control on non-convertible inputs—namely the professional staff—is a much more difficult problem, and brings into play the personnel function in the acquisition and maintenance of staff.

Customer influence

This summary of the approaches employed in the management of capacity, inventory and scheduling, conceals the fact that the management of the system has not always been totally successful. The Architects Department has in the past maintained its performance by employing the policy that when jobs are delayed due to site planning difficulties, etc, they are either speeded up by the application of managerial or political attention, or superceded by another job from the pipeline in order to maintain the requisite number of starts. This has, as would be expected, an undesirable effect on productivity though not profitability, because building costs have risen faster than salaries, and the fee is based on final cost. The department is now therefore committed to improving productivity and is encountering some obstacles particularly as regards the use of stocks of component drawing (e.g. type plans or even PDPs) and buying in systems. The reasons for such 'opposition' may be as follows. The total housing building programme system is essentially structure DQO—the tenant has to queue whilst the dwelling is designed and constructed for him or her. The tenant would certainly see it in this manner because for him the on-line aspects— the conversion process leading to the habitable dwelling—are most important. The politicians involved might see it this way, or as SQO because the process of providing many dwellings for many tenants ties up resources continuously. Either way however the total system involves a customer queue. For those concerned with the design of a dwelling, the customer in theory does not appear. The designer's customer is the next sub-system i.e. construction. Theoretically, the architect's customer is the builder who receives his drawings, whilst the builder's customer is the Housing Management Department who lets the dwelling. However as housing management and the housing development are organizationally and geographically in close proximity, there is in fact some feedback to those briefing the architect on receipt of the site. For this reason the architect will be getting from his client a paradigm of his ultimate customer, and he is asked to convert this brief into drawings for the builder. A similar situation might exist in a commercial organization i.e. the Housing Department represents marketing, servicing and sales, the Architects Department the design unit, and the builder the manufacturing/production unit. The man in the machine shop gets the drawing for a part, machines it and hands it on to the next man for, say, assembly with other components. The man employs skill, but his work is virtually totally prescribed by the designer's drawings, and *therefore* the need to make the part within the tolerances acceptable for the next part of the system. With the design unit, however, the position is different. Although the designer will receive a clear and detailed brief from the marketing section, his design will be informed by his own ideas, and his perception of the ultimate customer's needs. While in theory he is designing something for the man in the machine shop, he will also design with the ultimate customer in mind. Thus in the building development programme, whilst the ultimate customer is of little importance to the builder, he has a great effect on the management of the architect's work, as he will perceive the prospective tenants on the housing waiting list, whilst undertaking work that will be essentially qualitative and non-physical. He is therefore susceptible to customer influence in interpreting the client's brief into information for the builder.

Chapter 16

Limitations and Developments of Concepts

The Housing Building Programme case in Chapter 15 incorporates two features not specifically considered in the development of our concepts: firstly the absence for most part of a dominant physical flow; and secondly the extensive influence of the customer.

Excepting during construction, allocation and management much of the activity of the housing development system is focussed upon largely non-physical items, e.g. designs. Much of the flow prior to construction (stages 2–8 inclusive on Exhibit 15.3) comprises information, and although it was possible to some extent to model these sub-systems using our basic operating (i.e. physical resource) system structures, clearly the concepts must be further developed if they are to be of any substantial value in describing, comparing, and explaining such information processing systems. Whilst our concepts and our approach to the analysis of operating systems might be relevant in such cases, it is clear that, at least, we must develop some rules or guidelines for application.

It would be beneficial also if we were to develop rules or guidelines to take account of customer influence which may well be a significant factor in non-physical, and interpretative systems such as design operations. For example in seeking to describe or model such a system, we must ensure that:

(1) We recognize the existence of any customer 'brief', i.e. any feedback loop in the system in order that any activity caused or influenced by perception of the customer can be detected, hence
(2) We recognize the existence and location of the customer queue, and
(3) We recognize the type of customers that comprise the queue, in order that we identify the type of individual who may influence the operation of the system.

Ultimately, of course, all systems have some customer push sub-system, yet only in certain types of operation will the existence of such customers be of relevance in examining prior sub-systems. Clearly (1) above may be of relevance in private or public sector organizations. As for (3) above, it is unusual for private sector firms to know exactly who the customer will be. An architect working for a construction company will have to rely more on the marketing brief for Mr Average, than a council architect who will come face to face with tenants in the course of his work. 'Customer influence' will only occur in 'professional' type situations. A firm making and supplying prepacked Chinese dishes to take-away outlets will not perceive the queue of ultimate customers, nor will the customers influence the production of the dishes.

DEVELOPMENT OF CONCEPTS

In order to finish on a more positive note, this final section sketches out a few areas in which our simple concepts might be developed or extended. Time, space, and lack of imagination prevents a detailed discussion, hence for the present we simply offer six hypotheses: two ill-developed but hopefully not ill-conceived; two somewhat more robust; one entirely speculative; and one which will be developed elsewhere.

(1) *Firstly*, picking up a point from a previous section we must hypothesize that:

our concept of system structure(s) might usefully be developed to accommodate information as a resource and information flow as a principal feature of the system.

(2) *Secondly*, one might argue that to some limited extent:

the nature of the problems facing operations management in the other problem areas (i.e. other than capacity, scheduling and inventory) may be influenced by system structure. (For example the nature of the quality management problem—in particular the relative importance of inprocess quality control, and acceptance testing procedures—may be influenced by structure).

(3) *Organization structure and system structure* Our concepts concern the physical structure of systems. We have given little thought to organizational structure, yet it could be argued that the two must be connected. Organizational structure relates to the nature or shape of the organization as it might appear on an organizational chart; the manner in which the whole is organized into divisions and/or hierarchical levels, the authority structure, etc. It concerns the arrangement of individuals and their interrelationships and functions. In contrast, system structure, as discussed in the previous chapters, is concerned with the arrangement of the physical parts of the operations process(es).

Two arguments can be advanced in support of the hypothesis that organizational structure and systems structure must be related. Firstly—the classical argument—that the structure of an organization may in part reflect the nature of the technology employed. Since the systems structure may also be associated with technology, the two must bear some relationship. Secondly—a more recent argument—that since both organizational structure and system structure may be influenced by the amount of uncertainty faced by an organization, the two must bear some relationship. The latter argument appears to be the more convincing and is worth elaborating.

It has been argued[1] that the degree of uncertainty facing an organization influences the organizational structure which is either chosen or which evolves. As uncertainty increases, the amount of information to be processed in making decisions within the organization also increases, and hence the organization must take suitable steps in order that it might cope. Four organizational design strategies have been identified which, if employed, might:

(a) Reduce the need for information processing by:
    (i) The creation of slack, that is spare, resources, and/or
    (ii) The creation of self-contained tasks.
(b) Increase the ability of the organization to process information by:
    (i) The introduction of vertical information systems, and/or
    (ii) The creation of lateral relations.

The creation of slack resources reduces the number of exceptions that occur simply by means of reducing the required level of performance. The creation of self-contained tasks implies the establishment of groups with all resources necessary to perform the task, such groups affording increased flexibility, responsiveness, etc. The use of vertical information systems provides a supplementary mechanism for the processing of information and therefore avoids the overloading of conventional hierarchical communication channels. The creation of lateral relations enables the level of decision making to be moved down the organization so as to be closer to the point at which the information originates.

If we consider the level and nature of demand as one source of uncertainty upon an organization we may recognize a somewhat similar relationship between system structure and uncertainty. Increasing uncertainty in the levels of demand will influence the choice of capacity management strategy. The use of capacity management strategy 2a (i.e. the provision of excess capacity in order to reduce the need for capacity adjustment) is comparable to the creation of slack resources as a means of reducing the need for information processing. The adoption of other management strategies may also have organizational design implications. For example, the provision for effective capacity adjustment may require or depend upon the creation of effective lateral relations and vertical information systems. Certainly, improved information systems and lateral relations may facilitate the transfer of resources between operations. It is conceivable therefore that the employment of particular capacity management strategies as a means to deal with demand level uncertainties results in or requires the adoption of particular organizational design strategies, or vice versa.

Increasing uncertainty in the nature of demand will lead progressively to the abandonment firstly of output stocks and then input resource stocks. The use of certain systems structures as a means to deal with demand uncertainty may therefore correspond to the use of particular organizational design strategies. For example the use of self-contained tasks, the adoption of lateral relations, the existence of slack resources, and the use of vertical information systems may be more evident in structures SOD, DOD and SCO.

For these reasons it may be reasonable to hypothesize that the circumstances which give rise to the adoption of a particular system structure, and a particular strategy for the management of that system structure, also influence organizational structure. It is possible therefore that given a particular system structure we might expect to find, or require, a certain type of organization or vice versa.

(4) *Preferred systems structures and strategy.* Uncertainty in the environment, in particular uncertainty in the nature of demand, influences the feasibility of particular systems structures. Within such feasibility constraints operations management may have the opportunity of choosing a particular structure on the basis of desirability. We have given little thought to this question of desirability and must now consider the possibility of there being *preferred structures* which will be chosen within constraints imposed by feasibility. Similarly in identifying alternative strategies for the management of systems, in particular strategies for capacity management, we indicated that given a certain structure certain strategies might be available. Given the possibility of choice between available and feasible strategies we must consider the possibility of preference.

It is conceivable for example that structure SOS will be preferred to structure DOS, since SOS better insulates the operation within the system, that is the 'core' of the system, from environmental uncertainties. Similarly, it is conceivable that the use of a capacity management strategy which employs output stocks to minimize the need for capacity level changes will be preferred to a strategy of manipulating system capacity as a direct response to demand level changes. Organizational design theory again provides some corroborative evidence.

Thompson,[2] in attempting to develop theories for organizational design,

---

2.1
Under norms of rationality organizations seek to seal off their core technologies from environmental influences.

2.2
Under norms of rationality organizations seek to buffer environmental influences by surrounding their technical cores with input and output components.

2.3
Under norms of rationality organizations seek to smooth out input and output transactions.

2.4
Under norms of rationality organizations seek to anticipate and adapt to environmental changes which cannot be buffered or levelled.

2.5
When buffering, levelling and forecasting does not protect their technical cores from environmental fluctuations, organizations under norms of rationality resort to rationing.

Exhibit 16.1 Propositions for organizational design (Thompson[2])

presented five propositions which may be of some relevance here. They are listed in Exhibit 16.1. He is concerned with environmental influences on organizations (which for our purpose can be seen as those uncertainties referred to above, in particular uncertainties in demand level and the nature of demand) and the manner in which organizations are designed to accommodate these uncertainties. He refers to core technologies which for our purposes can be seen to be the key operations within operating systems. He attempts to develop propositions for rational behaviour in designing organizations. Summarizing the propositions in Exhibit 16.1. Thompson writes as follows:

'Perfection in technical rationality requires complete knowledge of cause/effect relations plus control over all of the relevant variables ... Therefore under norms of rationality organizations seek to seal off their core technologies from environmental influences. Since complete closure is impossible they seek to buffer environmental influences by surrounding their core technologies with input and output components. Because buffering does not handle all variations in an unsteady environment, organizations seek to smooth input and output transaction, and to anticipate and adapt to environmental changes which cannot be buffered or smoothed, and finally, when buffering, levelling and forecasting do not protect their technical cores from environmental fluctuations organizations resort to rationing.'

These are in effect the 'manoeuvering' devices which enable an organization to exercise some choice in control of the structure despite the dependence upon the environment and the affect of environmental uncertainties.

These propositions have perhaps some relevance in respect of choice and preference in both system structure and management strategy. Propositions 2.1 and 2.2, for example, support the hypothesis that given feasibility structure SOS will be preferred in the case of customer pull systems, and structure SQO in the case of customer push systems. Propositions 2.3, 2.4, and 2.5 may have some bearing upon preferred capacity management strategies. We might conclude for example that given feasibility the preferred strategy will be strategy 2bII, i.e. minimizing the need for demand level changes through the use of stocks. Failing or following the adoption of that strategy it might appear that the strategy 2bI will be employed.

In the event of the infeasibility of the adoption of these strategies it would appear that the organization will resort to manipulating capacity level, capacity levels having undertaken all possible action to eliminate or at least minimize the need for such changes.

Again, therefore, there exists some evidence to support the *hypothesis* that within feasibility constraints operations managers will consider certain structures to be more desirable than others and certain management strategies to be more appealing than others.

(5) *Structure life-cycles* Developing the logic of (4) above, is it perhaps conceivable that operating systems if successful will change their structure in some

predictable way as part of a 'maturing' process? Is there perhaps some 'life-cycle' pattern for systems? For example, in seeking to exploit a new market, an organization may establish a system with structure DQO. If this is successful it may be desirable, or even necessary, to change to structure SQO and then perhaps SCO. Take, for example, the provision of senior management courses. When first entering the business, the organization seeking to provide courses (say the 'college') will in effect provide structure DQO, i.e. they will try to identify customers, 'sign them up', and then arrange courses for them to suit their needs, using largely bought in or hired resources. If this type of operation is successful, i.e. if demand increases, it may be possible to employ some resour-ces, i.e. take on teaching staff, buy premises, equipment, etc. Furthermore, such a move may be desirable in order to meet increased demand and/or to provide an acceptable customer service by reducing waiting time. Eventually it may be possible to move to structure SCO where customer waiting is virtually eliminated, perhaps through the provision of standard courses on a frequent basis.

Alternatively, it is perhaps possible to identify a life-cycle for declining systems. Certainly there will be many situations in which systems move from SQO to DQO. Declining demand may necessitate the disposal of resources which are becoming underutilized. Within some organizations successful, growing systems may need to borrow resources from other systems in order to accommodate demand peaks (i.e. capacity management strategy I implemented through the transfer of resources between systems—see Exhibit 7.2), and in general resources will be traded from systems which have lower utilization and those in which changes in resource stocks have less effect on customer service or those which demand less priority in terms of customer service.

There are perhaps therefore some grounds for offering a final specula-tive hypothesis, namely that over time systems which are developing or declining will change their structures in a predictable fashion.

(6) *Business policy implications* It has been argued that operations management makes an inadequate contribution to *business policy* (or corporate strategy) decision-making, yet ironically such decisions have substantial implications for operations management. Little seems to have been written about operations policy, or the role of operations in making and implementing business policy decisions, which is surprising since so much business policy literature is based on product/market considerations.[3] It was not our intention explicitly to explore this area, yet some of what has been said is pertinent. The final three hypotheses above, for example, are clearly relevant. If some relationship exists between product/service market characteristics, system structure and organization structure, one dimension for operations policy is clear, since given a system and organization structure an enterprise is well equipped to deal with certain 'ex-ternal' conditions, *or* given certain external conditions a particular system and organization structure is appropriate. Taking a complementary policy per-spective it has been shown that enterprises develop or 'grow' through various

stages[4,5] from, for example, a single unit managed by a sole proprietor, to several semi-autonomous units each reporting to a corporate HQ. We have hypothesized a similar metamorphosis for operating system structures within enterprises.

Considering the influence of business policy decisions *on* operations management decisions and the contribution of operations management *to* business policy decisions, we hypothesize:

(a) that through the choice of product/service and market/demand characteristics business policy decisions determine the feasibility of operating system structures and thus have an *indirect* influence on operations management decisions. Additionally, operations management objectives will in part be determined by business policy which therefore exerts a *direct* influence on operations management's choice of both system structure and strategy.
(b) that operations management contributes to business policy decisions by providing information on either
  (i) the existing structure and strategies, or
  (ii) the structure and strategies appropriate for proposed product/service and market/demand characteristics.

SUMMARY OF CHAPTER 16

It is *hypothesized*:

(1) that the concepts can be extended to accommodate information as a resource.
(2) that to some limited extent other problem areas are related to system structure.
(3) that there is a relationship between system and organization structure.
(4) that there are preferred system structures and management strategies.
(5) that over time system structure will change in a predictable fashion.
(6) that business policy decisions influence operations management by determining the feasibility of system structures, and the nature of objectives, and that operations management contributes to business policy decisions by seeking to match structures and strategies to given product/service and market/demand characteristics, or vice versa.

REFERENCES

1. Galbraith, Jay, *Designing Complex Organisations*, Addison Wesley, 1973.
2. Thompson, James D., *Organisations in Action*, McGraw-Hill, 1967.
3. Ansoff, I. H., *Corporate Strategy*, Penguin, 1968.
4. Chandler, A. D., *Strategy and Structure*, MIT Press, 1962.
5. Salter, M. S., 'Stages of Corporate Development' in *Business Policy: Teaching and Research*, Taylor B. and MacMillan, K. (Eds) Bradford University Press, 1973.
6. Wild, R., *Policy for Operations Management* (in preparation).

Index